EXCAVATION AND SUPPORT FOR THE URBAN INFRASTRUCTURE

Papers prepared for sessions sponsored by
the Geotechnical Engineering Division of the
American Society of Civil Engineers
in conjunction with the
ASCE International Convention and Exposition
September 14 and 15, 1992
in New York, New York

EDITED BY
T.D. O'ROURKE AND
A.G. HOBELMAN

GEOTECHNICAL
SPECIAL PUBLICATION
NO. 33

Published by the
American Society of Civil Engineers
345 East 47th Street
New York, New York 10017-2398

ABSTRACT

This proceedings, *Excavation and Support for the Urban Infrastructure,* contains papers which focus on various aspects of excavations and ground support technologies for renewal of the urban and suburban environments. There are three major subject areas covered by this work: 1) Environmental, legal, and contractual issues, 2) case histories, and 3) design and construction considerations. Treatment of environmental, legal, and contractual aspects provides an overview of cross-cutting issues which set the parameters, constraints, and special features of infrastructure projects. The case histories present unique or unusual solutions to the support and maintenance of adjacent structures, and also summarize the effects of excavation on adjacent facilities. Papers pertaining to design and construction considerations focus on cut-and-cover excavation systems, tunneling, urban blasting, and grouting technologies.

Library of Congress Cataloging-in-Publication Data

Excavation and support for the urban infrastructure: papers presented for sessions sponsored by the Geotechnical Engineering Division of the American Society of Civil Engineers in conjunction with the ASCE International Convention and Exposition, September 14 and 15, 1992 in New York, New York/edited by T.D. O'Rourke and A.G. Hobelman.
    p. cm.—(Geotechnical special publication: no. 33)
  Includes indexes.
  ISBN 0-87262-906-6
    1. Excavation—Congresses. 2. Retaining walls—Congresses. 3. Municipal engineering—Congresses. I. O'Rourke, T. D. II. Hobelman, A. G. III. American Society of Civil Engineers. Geotechnical Engineering Division. IV. ASCE International Convention and Exposition (1992: New York, N.Y.) V. Series.
TA730.E93   1992
624.1'52—dc20   92-27782
                                                    CIP

The Society is not responsible for any statements made or opinions expressed in its publications.

Authorization to photocopy material for internal or personal use under circumstances not falling within the fair use provisions of the Copyright Act is granted by ASCE to libraries and other users registered with the Copyright Clearance Center (CCC) Transactional Reporting Service, provided that the base fee of $1.00 per article plus $.15 per page is paid directly to CCC, 27 Congress Street, Salem, MA 01970. The identification for ASCE Books is 0-87262/92. $1 + .15. Requests for special permission or bulk copying should be addressed to Reprints/Permissions Department.

Copyright © 1992 by the American Society of Civil Engineers,
All Rights Reserved.
Library of Congress Catalog Card No: 92-27782
ISBN 0-87262-906-6
Manufactured in the United States of America.

COVER PHOTOS

Upper left: Temporary excavation support for the State of Illinois Center, Chicago, IL (Courtesy of T.D. O'Rourke)

Right: Excavation for Embarcadero III, San Francisco, CA (Courtesy of Construction Photographers, Sausalito, CA)

PREFACE

In June, 1990 the Earth Retaining Structures Committee of the Geotechnical Engineering Division organized a conference at Cornell University on the Design and Performance of Earth Retaining Structures. The conference provided a major update of earth support technologies, many of which are vitally important for construction in congested urban environments. After the Cornell Conference, the Earth Retaining Structures Committee decided that it would be appropriate to focus on the role of excavations and ground support systems for renewal of the urban infrastructure.

Excavation and associated earth support for foundations, underground structures, and utilities are critical for a coherent and effective program of infrastructure renewal. Moreover, the crowded nature of cities and suburban areas in combination with many sensitive facilities, which can be affected adversely by adjacent construction, places special demands on site exploration and excavation procedures, and encourages innovative technologies for ground support, subsurface construction, and quality assurance. Because infrastructure renewal is one of the most important challenges facing the U.S. at the end of the 20th Century and because New York City is a focal point for important infrastructure engineering projects, it was both timely and fitting to develop this theme in specialty sessions for the 1992 ASCE Conference in New York City.

Three sessions were organized covering: 1) Environmental, Legal, and Contractual Issues, 2) Case Histories, and 3) Design and Construction Considerations. These three sessions are reflected in the three major subject areas of this publication. The aim of covering environmental, legal, and contractual aspects was to focus on the broad issues which set the parameters, constraints, and special features of building in the urban and suburban environments. The case histories were chosen to present unique or unusual solutions to the support and maintenance of adjacent urban structures and utilities, as well as to summarize the effects of excavations on adjacent facilities. The intention of covering design and construction considerations was to provide guidance for assessing the impact of construction on the urban environment, and conversely—the impact of the urban environment on construction.

It is the practice of the Geotechnical Engineering Division that each paper published in a special publication be peer reviewed for its technical content and quality. The standards of review are the same as those for the ASCE *Journal of Geotechnical Engineering*. Abstracts were received for 20 invited papers. Ultimately, 13 papers received favorable reviews and were accepted for publication.

It should be recognized that all papers in this volume are eligible for discussion in the *Journal of Geotechnical Engineering*. They are also eligible for any appropriate ASCE award.

Thanks are extended to the authors whose papers appear in this volume. Thanks also are extended to the 20 reviewers of papers in this volume, many of whom serve on the ASCE Earth Retaining Structures Committee. All reviewers were selected on the basis of their technical expertise in the subject area of the reviewed papers.

We are deeply grateful for the assistance of Laurie Mayes, who was of critical importance in coordinating paper submissions and reviews.

We also appreciate the excellent support from Shiela Menaker who arranged for the assembly and printing of this special publication.

T.D. O'Rourke	A.G. Hobelman
Professor	Vice President
Civil and Environmental Engineering	The George Hyman
Cornell University	Construction Company

CONTENTS

ENVIRONMENTAL, LEGAL, AND CONTRACTUAL ISSUES

Overview of Design and Construction in the Urban Environment
  Thomas R. Kuesel 1
Contracting and Legal Issues
  Robert A. Rubin and Jeannette L. Molina 6
Excavations and Contamination
  Bryan P. Sweeney and Joel S. Mooney 26
Oppportunities and Constraints for the Innovative Geotechnical Contractor
  Peter J. Nicholson and Donald A. Bruce 46

CASE HISTORIES

Limehouse Link Tunnel Project, London: A Case History
  Patrick McCreight, David Scott, and George Tamaro 65
The Reconstruction of the Morton Street Evacuation and Ventilation Shaft
  Daniel M. Hahn 91
Building Protection from Tunneling in Downtown Los Angeles
  Loring A. Wyllie, Jr. and John A. Dal Pino 107
Deep Cuts and Ground Movements in Chicago Clay
  Richard J. Finno 119

DESIGN AND CONSTRUCTION CONSIDERATIONS

Excavation and Support Systems in Urban Settings
  J.P. Gould, G.J. Tamaro, and J.P. Powers 144
Tunneling in the Urban Environment
  Norman A. Nadel 172
Frequency Based Control of Urban Blasting
  Charles H. Dowding 181
Construction Induced Vibration in Urban Environments
  Barry M. New 212
Grouting Techniques for Excavation Support
  Joseph P. Welsh 240

Subject Index 263

Author Index 264

GEOTECHNICAL SPECIAL PUBLICATIONS

1) TERZAGHI LECTURES
2) GEOTECHNICAL ASPECTS OF STIFF AND HARD CLAYS
3) LANDSLIDE DAMS: PROCESSES RISK, AND MITIGATION
4) TIEBACKS FOR BULKHEADS
5) SETTLEMENT OF SHALLOW FOUNDATION AND COHESIONLESS SOILS: DESIGN AND PERFORMANCE
6) USE OF IN SITU TESTS IN GEOTECHNICAL ENGINEERING
7) TIMBER BULKHEADS
8) FOUNDATIONS FOR TRANSMISSION LINE TOWERS
9) FOUNDATIONS AND EXCAVATIONS IN DECOMPOSED ROCK OF THE PIEDMONT PROVINCE
10) ENGINEERING ASPECTS OF SOIL EROSION DISPERSIVE CLAYS AND LOESS
11) DYNAMIC RESPONSE OF PILE FOUNDATIONS—EXPERIMENT, ANALYSIS AND OBSERVATION
12) SOIL IMPROVEMENT—A TEN YEAR UPDATE
13) GEOTECHNICAL PRACTICE FOR SOLID WASTE DISPOSAL '87
14) GEOTECHNICAL ASPECTS OF KARST TERRAINS
15) MEASURED PERFORMANCE SHALLOW FOUNDATIONS
16) SPECIAL TOPICS IN FOUNDATIONS
17) SOIL PROPERTIES EVALUATION FROM CENTRIFUGAL MODELS
18) GEOSYNTHETICS FOR SOIL IMPROVEMENT
19) MINE INDUCED SUBSIDENCE: EFFECTS ON ENGINEERED STRUCTURES
20) EARTHQUAKE ENGINEERING & SOIL DYNAMICS (II)
21) HYDRAULIC FILL STRUCTURES
22) FOUNDATION ENGINEERING: CURRENT PRINCIPLES AND PRACTICES
23) PREDICTED AND OBSERVED AXIAL BEHAVIOR OF PILES
24) RESILIENT MODULI OF SOILS: LABORATORY CONDITIONS
25) DESIGN AND PERFORMANCE OF EARTH RETAINING STRUCTURES
26) WASTE CONTAINMENT SYSTEMS: CONSTRUCTION, REGULATION, AND PERFORMANCE
27) GEOTECHNICAL ENGINEERING CONGRESS
28) DETECTION OF AND CONSTRUCTION AT THE SOIL/ROCK INTERFACE
29) RECENT ADVANCES IN INSTRUMENTATION, DATA ACQUISITION AND TESTING IN SOIL DYNAMICS
30) GROUTING, SOIL IMPROVEMENT AND GEOSYNTHETICS
31) STABILITY AND PERFORMANCE OF SLOPES AND EMBANKMENTS II (A 25-YEAR PERSPECTIVE)
32) EMBANKMENT DAMS - JAMES L. SHERARD CONTRIBUTIONS
33) EXCAVATION AND SUPPORT FOR THE URBAN INFRASTRUCTURE
34) PILES UNDER DYNAMIC LOADS

Overview of Design and Construction
in the Urban Environment

Thomas R. Kuesel[1]

Abstract

Urban underground construction is paralyzed by excessively cumbersome environmental and regulatory processes. Simplifying and expediting these systems could reduce cost and duration of construction. Protection of adjacent property, construction logistics, and disposal of excavated materials complicate urban projects, and have engendered special techniques to deal with them.

Introduction

At a recent underground construction conference I met an engineer who had just returned from a mine development project in Papua, New Guinea. The ore deposit was deep underground, so the first task was to sink a shaft. They brought in a shaft sinking rig and set about recruiting a local labor force. As the rig was being set up, the project manager was informed that before they started digging they would have to sacrifice a bullock. He airily dismissed this notion, and the labor force evaporated. To get them back, he had to agree to a full-fledged ceremony, complete with medicine man, flaming torches, whirling dancers, and when everyone was sufficiently worked up, the bullock ceremony. Everyone went home drunk, but the next morning they all showed up and turned to with a vim, and in a month the shaft was dug out.

Now this sounds like a quaint story of a "less developed society". But I wonder, how do we go about digging a shaft in New York City? First, we hold meetings to apply for authorization to study the project. Then we wait for budget approval for the study and assignment of a staff. Then we hire consultants to make an Environmental Impact

---

[1] Consulting Engineer, 5 Wood Lane, Charlottesville, VA 22901

Study. These include archaeologists, historians, water quality chemists, air pollution specialists, fish and wildlife conservators, endangered plant botanists, economists, and experts in legal and regulatory matters. We organize a Community Participation Program, which digs into housing patterns, social unrest, potential employment and housing dislocations, and the special concerns of every ethnic group on the voting rolls. If the project survives this phase, it goes into the political arena, where it competes for sponsors and against rival projects or other priorities for funding. This engenders more studies, reports, and meetings.

After perhaps two years and the expenditure of well over a million dollars, we may start discussion of the technical problems of digging the shaft. Eventually a contract is prepared, with five contract drawings related to the shaft and 55 concerned with traffic maintenance, utility relocations, protection of existing facilities, environmental control works, and project signs. The contract book is two inches thick, and contains a dozen pages related to actual shaft construction, with the rest devoted to requirements of government social programs, legal relations and regulatory requirements, and progress reporting and accounting specifications. The section on insurance is four times the length of the one on shaft construction.

I wonder what the Papuans would think of our process. Who has the more developed, enlightened, efficient society? I move that we strike Articles 1 through 63, and in lieu thereof, sacrifice one bullock.

The central problem of urban design and construction is paralysis. We are so encumbered with other agendas that we have lost sight of the original objective, which was to build something that would benefit society. So the most important item on our agenda is to bring to the attention of the general public, and particularly elected political and legislative leaders, the enormous cost and time burdens inherent in our bloated environmental, regulatory, and public works contracting processes.

We shall never match the beautiful efficiency of the Papuan medicine man, but surely we can do better than we are doing now. One important factor is that we have too many woodsmen chopping at individual trees, and no one looking at the forest. Each specialist is engaged to study his or her particular problem in isolation, and rarely do you find anyone with perspective, and a sense of where all the pieces fit and which ones are important. As in so many areas of design and construction (geotechnical design criteria furnish a glaring example), this problem is magnified by the liability syndrome. With no participation in how his report will be used, and mindful that some lawyer may one day wave it in his face, each specialist researches and documents his case ad nauseam and makes the most conservative recommendations imaginable. No one would dare to make a superficial study of an apparently superficial problem. It's like trying to sell Grade B milk, which might be more nutritious and appropriate, but no one will buy anything but Grade A.

We yearn for the medieval idea of the Architectus, the master builder who personally controlled every aspect of the building of a cathedral. Society will no longer entrust that responsibility to any one individual. The last to have such power was Robert Moses, whose accomplishments will be praised by future generations, but the present generation is more inclined to decry his disruption of existing society. What might be realistic, and certainly would be helpful, would be to designate and authorize a person or a small group to coordinate and expedite all the preconstruction studies, reports, and permits, to keep the process from bogging down.

A bit of personal experience may illuminate the matter. On the BART project in San Francisco, my job in the central office of the general design consultants was, basically, to make sure that nothing went out until it was right. In the office next to mine was Tom Rogers (incidentally, a former New York District Engineer for the Corps of Engineers), whose job was to see that everything went out on schedule, whether it was right or not. We understood each other perfectly, and got on very well. Tom would identify potential bottlenecks six months in advance, and I would assign a SWAT team to expedite the problem and make sure it was resolved before it became critical. It was a good system. There were very few nasty surprises resulting from something having fallen into a crack.

As for the regulatory system, I long ago evolved a theory about New York City, which might apply as well to other large cities. Over the generations the City has dealt with so many crooks and ne'er-do-wells that it no longer has the ability to deal with honest people. Each scandal has produced a new law or regulation designed to make sure that one doesn't happen again, until by now the clear presumption of the body of construction contracting law and regulations is that everyone who deals with the City is a potential crook. The preponderant emphasis is on the detection and punishment of transgression. We need to balance this with a system to reward accomplishment and excellence. The contracting method of Award Fees has been used on some projects for both design and construction, to reward the degree to which the Owner's stated objectives have been attained. In practice, it has rarely worked as well as it ought to, because in many cases the control of the environmental and regulatory processes is beyond the reach of either the engineer or the contractor. The only real power over these processes is that of the Sovereign, senior governmental authority, and this authority moves slowly because it is distracted by other agendas. Nonetheless, there is room here for some imaginative improvements in the regulatory and administrative process.

The common thread of these problems is a lack of understanding of the importance and value of time. Every engineer and contractor knows that a job that is delayed accumulates cost and loses direction. Few contract administrators, and no auditors, legislators, or general government officials seem to understand this. Everyone complains about the high cost of urban construction. Few recognize that the cost is not in doing the work, but in clearing out the thickets so that the contractor can get at the work, and in all the extraneous activities that are required in order to be permitted to work.

But let us try to look beyond our frustrations and assume that, miraculously, we are permitted to do actual design and construction work. The first thing that stands out about urban underground construction is that we are concerned less with what it is that we are trying to build, and more with protecting everything that everyone else has ever built near-by, and with maintaining public and private access and right-of-way around, above, and frequently through the work site. The second characteristic is that the contractor's access and right-of-way (or working space) are severely restricted, and the logistics of getting personnel, equipment and materials in and out frequently govern the work program and even the construction methods. And the third characteristic is the classic ecological dictum that nothing just goes away. The removal, transport and disposal of excavated materials so controls an urban underground construction project that unless the Owner works out at least a general plan, he is unlikely to receive bids. To these major problems you can add noise control, dust suppression, and sidewalk superintendents.

With this litany of distractions, it is a wonder that anything ever gets built at all. It is no surprise that urban construction has developed some very specialized techniques to cope with them while permitting construction to proceed. Some of the most ingenious are those for support of excavations, such as slurry wall and bored pile systems. Space restrictions led to tie-back anchors to

URBAN ENVIRONMENT OVERVIEW

eliminate interior cross-lot bracing. Recognition of the value of time (by developers and contractors) led to "top-down" construction.

As in all underground work, control of ground water is a major concern of both the engineer and the contractor. In an urban project, the technical problem of improvement of ground characteristics to permit efficient excavation is essential, but insufficient. The real problems are how to control settlement under adjacent properties, and how to clarify and detoxify the ground water so that it is clean enough to put into the city sewer system.

With such a myriad of interacting and conflicting considerations, virtually every large urban underground project is a unique situation, with its own special solution. Case histories are instructive and the sources for ideas on how to solve the next special situation, but it is rare that a full solution can be copied from one project to another. Nonetheless, many of the problems are generically common, and general techniques for handling them have evolved. We all learn from each other's adventures, and if we listen, we may learn something that will make our own adventures less exciting.

To this end, the remainder of this afternoon's session will illuminate the murky areas of environmental and regulatory issues. Tomorrow morning will provide a sampling of case histories, and tomorrow afternoon a look at some innovative design and construction techniques that may make your lives easier. Heads up! Ears Open! And may we all go away wiser than we came.

EXCAVATION AND SUPPORT FOR THE URBAN INFRASTRUCTURE: ENVIRONMENTAL AND REGULATORY ISSUES

CONTRACTING AND LEGAL ISSUES

Robert A. Rubin, Esq.[*]
Jeannette L. Molina[**]

Abstract: Urban excavation is fraught with risks and potential liability for environmental and regulatory issues. Federal law, state law and local regulations all must be reviewed for their potential effect on a particular project. Preliminary requirements must be satisfied prior to the commencement of excavation, including compliance with the National Environmental Policy Act. Engineers and contractors face liability for uncovering hazardous materials under the Comprehensive Environmental Response Compensation and Liability Act (C.E.R.C.L.A.). They also face liability to public utility companies and adjoining property owners under the "doctrine of lateral support" as well as for interference with business interests. Theories of liability range from those imposed by statute or regulation to common law liability based on negligence or strict liability.

I. Introduction

An increasing concern among contractors and engineers who engage in construction in urban environments is the potential for being exposed to liability for causing or creating environmental contamination or increasing risks to health and safety. The concerns must be dealt with at the early stages of

---

[*]Robert A. Rubin, F.ASCE, is a partner in the New York City law firm of Postner & Rubin, whose practice is devoted to construction matters.

[**]Jeannette L. Molina is a second-year law student at Hofstra University.

construction. This paper addresses the legal aspects of environmental and regulatory issues which arise during the excavation phase of urban construction.

The issues addressed cover a broad range of topics, including environmental regulation requirements and common law bases for imposing liability. The first topics discussed deal with the preliminary requirements which must be satisfied prior to commencement of excavation. These preliminary requirements include compliance with the National Environmental Policy Act, which requires the filing of an environmental impact statement for any major federal action undertaken which may affect the environment. Further, engineers and contractors may be held liable for uncovering hazardous materials under the Comprehensive Environmental Response Compensation and Liability Act. In addition, engineers and contractors must comply with state and federal regulations which are modelled after the federal Occupational Safety and Hazardous Act, the Clean Water Act, and the Federal Safe Drinking Water Act.

In addition to environmental regulations, common law doctrines which impose liability regarding duties owed to public utility companies by those excavating in urban environments, include theories of negligence, strict liability or trespass. Not only are common-law doctrines applicable, but state statutes and ordinances must be adhered to by engineers and contractors in order to avoid liability. Also discussed is the "doctrine of lateral support", which is derived from common-law principles and is also incorporated into state statutes and ordinances. The doctrine of lateral support encompasses theories of liability which range from negligence to strict liability. Private excavators and engineers can be held liable under this doctrine as well as governmental entities working in that capacity.

Other common-law based causes of action include liability for noise created during excavation, interference with business interests and business interruption where excavation is taking place on adjacent properties. Finally, the permitting process is briefly addressed as an indication of an additional requirement imposed which makes excavation projects difficult if not impossible to complete successfully.

II. The National Environmental Policy Act.

A. The Environmental Impact Statement.

One of the first major enactments regarding environmental liabilities was the National Environmental Policy Act ("NEPA").[1] NEPA's requirement that an Environmental Impact Statement ("EIS") be prepared for all major construction projects undertaken or financed by the federal government led to a change in the construction industry. It enhanced the awareness and increased the liabilities of contractors and engineers resulting from effects of construction projects on the human environment.[2]

The EIS must be filed for every proposal of a major federal action affecting the human environment. It must contain extensive data furnished by the contractor or engineer applying for a permit or other major action being considered by the government agency.[3] A preliminary determination must be made as to whether the project constitutes "a major federal action that would significantly affect the quality of the human environment."[4]

B. **Standing To Sue.**

After a preliminary determination has been made, a determination must be made as to whether an individual has standing to sue. Private citizens, associations or organizations may maintain an action in federal court for injunctive relief against commercial development or activities alleged to be harmful to the environment.[5] The test a plaintiff must satisfy to have standing to sue in federal court to protect the environment from harm is: 1) that the challenged action has caused him injury in fact; and 2) the interests which he has asserted are within the zone of interests sought to be protected by the statute or constitutional guaranty in question.[6] The interest sought to be protected may reflect "aesthetic, conservational, and recreational" values.[7]

This determination is made with respect to all phases of construction, including the environmental effects caused by excavation.[8] Once the determination is made that the project significantly affects the quality of the environment, the standard by which the EIS is reviewed is "that the appropriate federal agency has gone through a process of individualized consideration and balancing of factors -- conducted fully and in good faith."[9] This entire process is only preliminary regulation imposed on contractors and engineers during the excavation phase of construction in federal projects. It requires the most fundamental compliance with regulations to avoid liability and does not absolutely absolve contractors and engineers from liability. Where private projects are concerned, state regulations similar to NEPA are imposed.

III. **The Comprehensive Environmental Response Compensation and Liability Act.**

Other liabilities may arise during the excavation phase of construction involving hazardous wastes in the environment released or uncovered during excavation. Exposure to contractors or engineers for liabilities arises from hazardous wastes, PCBs, asbestos and other hazardous materials.[10] Under the Comprehensive Environmental Response Compensation and Liability Act ("CERCLA")[11] and its amendments entitled the Superfund Amendment and Reauthorization Act ("SARA")[12],

> any person is jointly and severely liable for the cost of cleanup of a hazardous substance that has been deposited, stored, disposed of, place, or otherwise located in a building structure, installation or any other known location. The five classes of persons subject to such liability are:
>
> 1. the owner and operator of a vessel;
> 2. the owner and operator of a facility;
> 3. anyone who at the time of disposal owned or operated the facility in which the hazardous substance were disposed;
> 4. anyone who arranged for the transport, disposal or treatment of hazardous substances; and
> 5. any person who accepts hazardous substances for transport to a site selected by such person.[13]

The result of this broad definition is that contractors and engineers who encounter hazardous materials can become generators of waste when it is moved, disturbed, hauled or disposed of during excavation.[14]

Other regulations which expose contractors and engineers to potential liability are the Occupational Safety and Hazards Act ("OSHA")[15], the Clean Air Act[16], the Clean Water Act[17], and the Federal Safe Drinking Water Act[18]. Individual states also have applicable state laws that apply to cleaning up hazardous wastes found and uncovered during excavation at construction sites.

IV. Liability Predicated on Negligence.

Liability of one excavating in the highway or street has often been predicted on the theory of negligence.[19] These cases deal specifically with injury to public utility cables, conduits or the like. Case law is consistent in permitting recovery where the contractor has failed to exercise reasonable care[20] and denying recovery where the contractor's negligence has not been proved.[21] For example, in Southwestern Bell Tel. Co. v. Rawlings Mfg. Co.[22], the court held that the plaintiff telephone company was entitled to recover for damages sustained when its underground cables in a public alley were caused to fall into an excavation made by one of the defendants. The court held the defendant to a high standard, stating that the defendant knew from his experience in excavation work that there were underground installations of various utilities in the alleys and streets of the city. He knew the telephone conduit was exposed for about three weeks before it caved in. Additionally, he was the one who decided whether or not shoring and bracing should be done. He was not instructed by anyone in relation to shoring or bracing the exposed telephone conduit. Therefore, the defendant is liable by reason of his negligence in failing to take appropriate action based on the information available to him.

In another case, recovery was denied where negligence was not found. In Washington Gas Light Co. v. George A. Fuller Co.[23], there was a cave-in of the east side of an excavation in a street, which caused damage to plaintiff's gas main located in the bed of the street. The Court found for the defendant, concluding that the plaintiff had failed to prove by a preponderance of the evidence, that either the general contractor or the shoring subcontractor was negligent or that either had proximately caused damage to its gas main.

Some cases have even found that plaintiff's recovery is precluded if there is a showing of the contributory negligence. In Mountain States Tel. & Tel. Co. v. Horn Tower Construction Co.[24], the court sustained a finding that the plaintiff was guilty of contributory negligence where the defendant subcontractor, engaged in rough-grading a city street, struck and severed a large buried conduit and cable of the plaintiff telephone company. The court found that the cable lines were not located in compliance with the map presented to the city and that although plaintiff had knowledge of the street work to be done, it elected to run the hazard of having the street work with the conduit in the ground and was therefore contributorily negligent.

V. **Liability Based on Strict Liability or Trespass.**

Notwithstanding the negligence standard of liability in excavation, some courts have allowed for recovery on the basis of strict liability[25] or trespass.[26] An example of a case where strict liability was imposed is the Frontier Tel. Co. v. Hepp[27] case. Here, the court held that when an excavator uses the public street of a city for his own private purposes, and goes beneath the street surface, the duty rests upon him to fully inform himself as to what lies below, so that he may avoid injury to the property of the city or others which is rightfully theirs. This is based on the premise that the streets of a modern city are so underlaid with pipes and conduits of various kinds necessary to the comfort and welfare of its citizens that one may be required to almost take judicial notice that in digging he may encounter such pipes or conduit at any point in a street.[28]

An example of a trespass case is Illinois Bell Telephone Co. v. Deliso Construction Co.[29]. Here, the court held that when the excavating hoe of the defendant contractor struck and damaged the underground cable of plaintiff telephone company, the defendant committed a trespass for which it was liable.

VI. **Violations of Statutes or Ordinances.**

Further liability can be imposed for violations of statutes or ordinances enacted by the jurisdiction in which work is being performed.[30] For example, in Consolidated Edison Co. v. TJN Construction Co.[31], a New York City ordinance provided that notice of any excavation by a private party be given to any corporation whose pipes, mains, or conduits are laid in the street about to be disturbed by such excavation, at least 24 hours before commencement of excavation. The ordinance also imposed a duty on the excavator to protect the property of the utility. The court held that the defendant construction company was not liable for damages sustained by the plaintiff because defendant had given timely notice, sufficient to afford plaintiff an opportunity to advise the excavator of the location of subsurface lines to be protected. The court stated that the ordinance could not be read as imposing an absolute duty of protection at peril and that where defendant did not have actual or constructive notice of the location of the utility's subsurface structures, no liability in negligence could be found.

VII.   **The Lateral Support Doctrine.**

   A.   **Definition.**

Contractors and engineers may also be liable to adjacent landowners for excavation that threatens the physical integrity of adjacent land. Property owners are protected "from neighbor's actions that cause subsidence, erosion, or similar injury under two closely related common-law doctrines known as lateral support and subjacent support."[32] The lateral support doctrine relates specifically to excavation in urban environments and restrains excavation that would cause the surface of adjacent land to fall, crumble, slip, slide, erode or otherwise subside. This duty extends only to support the neighboring land in its raw or natural condition[33] and not to man-made structures placed upon the land.[34] However, most jurisdictions have extended protection to buildings and improvements where the damage to improvements is a consequence of withdrawal of lateral support which would have caused subsidence of land in its natural condition.[35]

   B.   **Liability Based on Negligence.**

Liability can also be imposed on the basis of negligence of the supporting landowner. In such cases, the minimum requirement is that the supporting landowner or excavator notify the owner of any excavation in advance of any work being performed.[36] Further liability based on negligence can be imposed when the excavator is aware or should be aware of a substantial risk of damage.[37]

   C.   **Governmental Entities.**

This liability extends not only to private individuals, but also to governmental entities involved in construction projects which withdraw support from privately owned property.[38] In many instances, governmental immunity is raised as a defense but it often fails as a bar to liability.[39] For example, the city of Providence was held liable in <u>Prete v. Cray</u>[40]. In this case, a city excavation to repair sewers beneath city streets caused quicksand from under adjacent owner's land to flow into the excavation, resulting in subsidence damage to adjacent owner's land and improvements. The court, finding liability, held the city to the same standards applicable to private excavators and owners because the municipality was not exercising a governmental function delegated to it by the state.[41] Under Prete, liability was predicated upon the nature of the governmental activity that removes support.[42]

D.   State Statutes and Ordinances.

As a result of the ambiguity created by inconsistencies in case law regarding the doctrine of lateral support, many states have adopted their own statutes and ordinances which modify the common-law lateral support doctrine. This legislation can generally be divided into four groups:

> 1. strict liability for maintaining lateral support is extended to improvements, so that the excavator is always responsible for supporting neighboring buildings and structures;
>
> 2. the owner of neighboring buildings and structures is responsible for their support, provided that the excavator is not negligent;
>
> 2. the responsibility for supporting neighboring buildings and structures depends upon the depth of the foundation of the neighboring building or other structure, assigning the duty to the excavator if the depth exceeds a statutory standard;
>
> 3. the responsibility for supporting neighboring buildings and structures depends upon the depth of the intended excavation, assigning the duty to the excavator if the depth exceeds a statutory standard.[43]

In most states, statutes set out minimum requirements, one of which is that the excavator shall notify the adjacent owner of building and structures of the proposed excavation. Where notice is also accompanied by a duty to protect the land, a permit or license may be required to enter the land where the improvements or structures are located in order to provide protection.[44]

E.   Ground Water Loss and Lateral Support.

The doctrine of lateral support is often relied on in cases involving settlement of structures resting on wooden piles that have decayed as a result of ground water loss, which usually happens in coastal cities.[45]

Subsurface wooden piles do not depend on the natural presence of water pressure beneath the land; rather, wooden piles depend on water's chemical properties to prevent decay.[46]

A landowner has the right to rely on support and ground water law to protect its artificial support system and prevent its decay. Support law, such as the doctrine of lateral support, provides for protection by holding defendant excavators who withdraw ground water and alter water tables either to a negligence standard of due care or to a strict liability standard with a prerequisite determination of whether the improvements on the land contributed to their own injury. For example, in Bjorvatn v. Pacific Mechanical Construction, Inc.[47], the court held the construction company strictly liable for lowering the water table in the process of excavating for a city sewer and, as a result, removing support from under the property owner's house. Because the digging of a deep trench deprived the compressible solid of its natural water content, the property owner's land sank, causing cracks in the foundation and walls of his house.[48] The court mandated recovery on a strict liability theory because the harm occurred in the exercise of the power of eminent domain.

VIII. Noise Created During Excavation.

Beyond these statutory or regulatory requirements, contractors and engineers have also been held liable to individuals bringing actions for trespass, nuisance or strict liability for noise created during excavation. In an often cited case, Celebrity Studios v. Civetta Excavating Inc.[49], a tenant of a rehearsal studio business in a building adjacent to an excavation site brought an action against the adjacent excavator for noise created by the excavators during the course of excavating and constructing an adjacent new building. The tenant sued on the grounds that the noise created during excavation "so impinged upon and impaired the operation of the rehearsal studio business as to warrant recovery."[50]

The action for recovery was premises on several theories. The first cause of action was based on the theory of strict liability, alleging that the blasting operations and the driving of piles on the adjacent property deprived the tenant of a quiet atmosphere and tranquility which was essential to the proper functioning of its business. The tenant argued that the excavating company was absolutely liable for damage caused by excessive noise and relied on an older case, Spano v. Perini[51], a blasting case where the court imposed liability. In Spano, the court declared that: "[i]n our view the time has come for this court to make the 'announcement' and declare that one who engages in blasting must assume (the) responsibility, and be liable without fault, for any injury he causes to neighboring property."[52] However, Spano involved physical damage to property for which recovery was allowed. The court in Civetta distinguished actual damage to property from intangible damage to silence and tranquility which cannot be quantified, and rejected the tenant's claim of absolute liability on the ground that serenity cannot be expected in an urban city such as New York City.

The second cause of action was based on the theory of negligence, alleging that the excessive noises and vibration during excavation was caused by the "neglect and failure of the excavating company to sue alternative methods and equipment to minimize the volume and intensity of the noise and vibrations,"[53] despite the tenant's complaints. This cause of action was dismissed on the ground that the duty of the excavator does not extend to the use of the best equipment, only that the excavator take reasonable precautions to prevent foreseeable harm.

The third cause of action was based on the theory of nuisance, in that the excavation and foundation laying and construction of a 47-floor office building is a continuous project which lasts for a long period of time, sufficient to warrant recovery. The court also rejected this theory on the notion that most case law recognized that "the temporary annoyances and inconveniences of construction cannot be considered under the law as nuisances which permanently deprive adjoining land of some of its value."[54]

The last cause of action was premises on the theory of trespass, alleging that the noise travelled upon the tenant's premises and was therefore a trespass. This cause of action was also dismissed on the ground that an invasion of sound waves does not cause tangible damages and does not constitute an occupancy of space.[55]

IX. Business Interruption Damages.

A cause of action has also been allowed where a business interest of a tenant is interfered with by an excavator. In Thornton v. Connelly[56], the owner of property on which a residence, a storehouse and other buildings were located, and where retail mercantile business was conducted such as was necessary for the use and convenience of the home and storehouse, brought an action against an excavator. The owner stated that the excavation, foundations and erection of a filling station, as contemplated by the excavator, were invasions of the public road. The court held that the owner of the property had a greater interest in a road or street on which the property abutted, both in kind and in degree, than the general public and that he was therefore entitled to maintain a suit to enjoin the obstruction of such road or street.

Excavating contractors and engineers have also been sued in tort for liability to a business for actions which give rise to economic damage due to business interruption that is unaccompanied by physical injury to the owner or property.[57] In such cases, the excavator or engineer being sued will only be held liable for injuries proximately caused by the engineer's or contractor's conduct. Therefore, recovery for business interruption depends on the nature of the act alleged to have caused the injury.[58]

For example, a contractor in the course of excavating a building site severed wires supplying electrical power to a printing plant. The court in Byrd v. English[59] did not hold the contractor liable for economic damage incurred by the printer plaintiff as a result of business interruption. The court stated that the printer suffered no personal injury or property damage; rather, the contractor caused damage to the property of the power company, which held a contract with the printer to furnish him electricity. The court ruled that the printer's contract granting him a right to electricity was with the power company and was not even remotely connected to the contractor. However, the court held that while the printer seeks recovery against the contractor to recover for injuries to his business, the power company might recover on account of the contractor's negligence, including any sums which the power company was compelled to pay in damages to its customers.[60]

Similarly, excavations for subway construction or repair work have been held not to give rise to compensable injuries under general principles of eminent domain or particular statutes which provide compensation to property owners injured by such projects.[61] Temporary interference with access to property as a result of such construction projects carried out by governmental or municipal agencies has been held to be not sufficient for recovery.[62] For example, in Downside Risk Inc. v. Metropolitan Atlanta Rapid Transit Authority[63], the court held that although the construction of a subway resulted in temporary interferences (e.g., smoke, dust, fumes, odors, noise and vibrations) to adjacent property owners, temporary obstruction of a right of ingress and egress does not deprive one of his private property.

In a similar case, Farrell v. Rose[64], a contractor retained by the city to construct a retaining wall along railroad tracks was held not to be liable to the owner of parking garages because the contractor's excavations and equipment had blocked the street on which the garages were located. The court stated that a property owner must bear the burden of any inconveniences and damages that may result from the temporary obstructions of repairing and maintaining public utilities. The damages are not considered compensable since the benefit from the improvement ultimately accrues to the owner.[65] However, the court recognized that where a contractor prolongs work unnecessarily or unreasonably, an owner has a right to be afforded relief.[66]

X. The Permitting Process.

The permitting process must also be considered by engineers and contractors before commencing any excavation for construction or repair of structures. Excavation, in itself, is considered a hazard to health and safety. To the extent that excavation must be regulated if it is considered a health hazard, it has been held that "the grant of power in a city charter to regulate the use of the streets implies the power to impose all such reasonable conditions, in relations to their use, as will tend to the accomplishment of the municipal duty to provide for the general welfare and safety of the community, in that respect."[67] The power to regulate conduct of those who carry on excavations of real property in cities usually involves purchasing a permit for excavation and prescribing punishment by fine and imprisonment where excavation is performed without a permit.[68] In addition to the permit requirement, a city may also have the "power to prescribe the mode and manner of the construction of building within city limits in the interest of health and safety."[69]

Conclusion

Urban excavation is fraught with risks and potential liability for environmental and regulatory issues.

Federal law, state law and local regulations all must be reviewed for their potential effect on a particular project.

ENDNOTES

1. 42 U.S.C. § 4332 et seq. The relevant portion of the statute applied here is as follows:

> The Congress authorizes and directs that, to the fullest extent possible: (1) the policies, regulations, and public laws of the United States shall be interpreted and administered in accordance with the policies set forth in this chapter, and (2) all agencies of the Federal Government shall- (A) utilize a systematic, interdisciplinary approach which will insure the integrated use of the natural and social sciences and the environmental design arts in planning and in decision-making which may have an impact on man's environment;
> (B) identify and develop methods and procedures, in consultation with the Council on Environmental Quality established by subchapter II of this chapter, which will insure that presently unquantified environmental amenities and values may be given appropriate consideration in decision-making along with economic and technical consideration;
> (C) include in every recommendation or report on proposals for legislation and other major Federal actions significantly affecting the quality of the human environment, a detailed statement by the responsible official on -- (i) the environmental impact of the proposed action, (ii) any adverse environmental effects which cannot be avoided should the proposal be implemented, (iii) alternatives to the proposed action, (iv) the relationship between local short-term uses of man's environmental and the maintenance and enhancement of long-term productivity , and (v) and irreversible and irretrievable commitments of resources which would be involved in the proposed action should it be implemented. Prior to making any detailed statement, the responsible Federal official shall consult with and obtain the comments of any Federal agency which has jurisdiction by law or special expertise with respect to any environmental impact involved. Copies of such statement and the comments and views of the appropriate Federal, State, and local agencies, which are authorized to development and enforce environmental standards, shall be made available to the President, the Council on Environmental Quality and to the public as provided by section 552 of Title 5, and shall accompany the proposal through the existing agency review processes;

(D) study, develop, and describe appropriate alternatives to recommended courses of action in any proposal which involved unresolved conflicts concerning alternative uses of available resources.

2. Bickers, M.S. (1988). *Environmental Liability in Project Construction and Management*, Society of Mining Engineers, Littleton, CO.

3. Shea, E.E. (1991), *Introduction to U.S. Environmental Laws*, unpublished.

4. Continental Illinois National Bank and Trust Co. of Chicago v. Kleindienst, 382 F. Supp 107, 110 (N.D. Ill. 1973).

5. 11 ALR Fed 556, 559.

6. Id. at 561.

7. Association of Data Processing Service Organizations Inc. v. Camp, 397 U.S. 150 (1970).

8. Friends of the Earth v. Coleman, 513 F.2d 95 (Ca. Cal. 1975), where the court held that the test for determining whether proposed excavations from a canal, proposed as part of a state water project, to obtain fill for a federal highway project, constituted a sufficiently significant commencement of a canal project to warrant formal evaluation of the canal's environmental impact did not depend upon interrelation of projects per se, but depended on whether completion of one project would inevitably involve an irreversible and irretrievable commitment of resources to the second.

9. The East 63rd St. Assoc. v. Coleman, 414 F. Supp. 1318 (S.D.N.Y. 1976), where the plaintiffs alleged that 1) the EIS did not describe the extent of excavation required for the 63rd Street Station or the environmental impacts of such excavation and other related construction; and that, 2) the EIS did not mention, much less analyze, the resulting visual blight, traffic congestion, air and noise pollution and similar impacts upon residents of 63rd Street; the court denied plaintiff's motion to enjoin further excavation and construction of subway station.

10. Estes, J. M. and Connolly, M.A., (Aug. 1990). "Avoiding Exposure to Environmental Liabilities: Concerns for Sureties," 10 The Construction Lawyer 33 American Bar Association.

11. 42 U.S.C. §§ 9601-9675.

12. 42 U.S.C. §§ 9619 et seq.

13. Id.

14. Most states have passed local Superfund statutes like the federal Superfund provisions.

15. 29. U.S.C. §§ 654 et seq.

16. 42 U.S.C. § 7401 et seq.

17. 33 U.S.C. § 1251 et seq.

18. 42 U.S.C. §§ 300 et seq.

19. 73 ALR3d 987, 990.

20. See, e.g., Edison Illuminating Co. v. Misch, 200 Mich. 114, 116 N.W. 944 (1918); Willmar Gas Co. v. Duinck, 236 Minn. 499, 53 N.W.2d 225 (1950).

21. See, e.g., Southwestern Bell Tel. Co. v. Rawlings Mfg. Co., 359 S.W.2d 393 (Mo. App. 1962); Mountain States Tel. & Tel. Co. v. Horn Tower Construction Co., 147 Colo. 166, 363 P.2d 175 (1961); Wash. Gas Light Co. v. George A. Fuller Co., 257 A.2d 498 (Dist. Colo. App. 1969); United Electric Light Co. v. Deliso Constr. Co., 315 Mass 313, 52 N.E.2d 553 (1943).

22. 359 S.W.2d 393 (Mo. App. 1963).

23. 257 A.2d 498 (Dist. Colo. App. 1969).

24. 147 Colo 166, 363 P.2d 175 (1961).

25. See, New York Steam Co. v. Foundation Co., 195 N.Y. 43, 87 N.E. 765 (1909); Frontier Telephone Co. v. Hepp, 66 Misc. 265, 121 N.Y.S. 460 (1910); Cincinnati and Suburban Bell Telephone Co. v. Eadler, 75 Ohio App. 258, 61 N.E.2d 795 (1944); Pioneer Natural Gas Co. v. K & M Paving Co., 374 S.W.2d 214 (Tex. 1963).

26. See, Illinois Bell Telephone Co. v. Foundation Co., 3 Ill. App.2d 258, 121 N.E.2d 600 (1954); United Electric Co. v. Deliso Construction Co., 315 Mass. 313, 52 N.E.2d

553 (1943).

27. 121 N.Y.S. 460, 66 Misc. 265 (1910).

28. Id.

29. 3 Ill. App.2d 258, 121 N.E.2d 600 (1954).

30. See, El Paso Electric Co. v. Safeway Stores, Inc., 257 S.W.2d 502 (Tex. Civ. App. 1953); Consolidated Edison Co. v. TNJ Construction Co., 52 Misc.2d 788, 276 N.Y.S.2d 979 (1967).

31. 52 Misc.2d 788, 276 N.Y.S.2d 979 (1967).

32. Smith, J.C. (1988). "Support of Land and Improvement," Neighboring Property Owners, Hand, J.P., Smith, J.C., Shepards-McGraw-Hill.

33. See, Gladin v. Von Engeln, 195 Colo. 88, 575 P.2d 418 (1978); First National Bank & Trust Co. v. Universal Mortgage and Realty Trust, 38 Ill. App. 3d 345, 347 N.E.2d 198 (1976).

34. Transportation Co. v. Chicago, 99 U.S. 635, 645 (1878).

35. See, e.g., Gladin v. Von Engeln, supra; Riley v. Continuous Rail Joint Co., 110 A.D. 787, 97 N.Y.S. 293 (1906), aff'd, 1983 N.Y. 643, 86 N.E. 1132 (1908); Kelly v. Falangus, 63 Wash.2d 581, 98 P.2d 223 (1964); Williams v. Southern Ry, 396 S.W.2d 98 (Tenn. Ct. App. 1965); Simons v. Tri-State Construction Co., 33 Wash. App. 315, 655 P.2d 703 (1983); Noone v. Price, 298 S.E.2d 218 (W. Va. 1982).

36. Schmidt v. Chapman, 26 Wis.2d 11, 131 N.E.2d 689 (1964); City of LaCrosse v. Jiracek, Cos., 108 Wis.2d 684, 324 N.W.2d 440 (Ct. App. 1982); Exchange National Bank v. Code, 23 Ill. App. 2d 382, 163 N.E.2d 554 (1959).

37. Beaver v. Hitchcock, 151 W.Va. 620, 153 S.E.2d 886 (1967); Lee v. Talao Building Development Co., 175 Cal. App.3d 565, 220 Cal. Rptr. 782 (1985); Pricket v. Sullivan, 190 Cal. App.2d 489, 12 Cal. Rptr. 55 (1961).

38. See, note 33, supra, at p. 95.

LEGAL ISSUES

39. See, e.g., Waters v. Biesecker, 309 N.C. 165, 305 S.E.2d 539 (1983), (the city alcoholic beverage control board was held liable for negligent failure to notify adjoining landowner of proposed excavation); Simons v. Tri-State Construction Co., 33 Wash. App 315, 655 P.2d 703 (1983) (action dismissed against city's contractor).

40. 49 R.I. 208, 141 A. 608 (1928).

41. Id. at 211.

42. Id.

43. See, note 39, supra, at p. 100.

44. Id. at 101. See, Heerey v. Berke, 179 Ill. App.3d 925, 534 N.E.2d 1277 (1989), (the court rejected an adjoining owner's argument that excavator's notice requesting license to enter the adjoining owner's property to protect it was inadequate because it failed to state the depth of proposed excavation).

See also, the National Building Code, § 904.2(a)(b)(1967), requiring that a deep excavator protect nearby buildings and structures, provides as follows:

> When an excavation extends not more than 10 feet below the curb level nearest the point of excavation under consideration, or below the surface of the ground where there is no such curb level, the owner of a building or structure adjacent to the excavation, the safety of which may be affected by such excavation, shall be notified in writing by the one causing the excavation to be made at least one week before the excavation is commenced. The owner of the adjoining structure shall preserve and protect the same from injury and, when necessary, shall pin and support the same by proper foundations. For such purpose, he shall be permitted, if necessary, to enter upon the premises where such excavation is being made.
>
> When an excavation extends more than 10 feet below curb level nearest the point of excavation under consideration, or below the surface of the ground where there is no such curb level, the person causing such excavation to be made shall, if afforded the necessary consent to enter upon the adjoining land, at his own expense, preserve and protect from injury every building or structure, the safety of which may be affected by such excavation and, when necessary, irrespective of the depth to

which the foundations of such buildings or structure may extend. If the necessary consent is not accorded to the person making the excavation, then it shall be the duty of the person refusing such consent to preserve and protect such building or structure from injury and, when necessary, to underpin and support the same by proper foundations; and that for that purpose such person shall, when necessary, be permitted to enter upon the premises where such excavation is being made.

45. Kincaid, Susan M. (1985). "Cities Supported by Sticks in the Mud: A Variation on the Settlement of Land and Structures Caused by Ground Water Removal," 15 BC Environmental Aff. L. Rev. 349, 382.

46. Id. at 383.

47. 77 Wash.2d 563, 464 P.2d 432 (1970).

48. Id. at 546.

49. 340 N.Y.S.2d 694, 72 Misc.2d 1077 (1973).

50. Id. at 695.

51. 25 N.Y.2d 11, 302 N.Y.S.2d 527, 250 N.E.2d 31 (1969).

52. Id. at 15.

53. Id.

54. Id. at 700; Dixon v. New York Trap Rock Corp., 293 N.Y. 509, 513, 58 N.E.2d 517; Maltbie v. Bolting, 6 Misc. 339, 26 N.Y.S. 903.

55. Id. at 703.

56. 15. Tenn. App. 436 (1936).

57. "Business Interruption, Without Physical Damage, as Actionable", 65 ALR4th 1126, 1128.

58. Id. at 1130.

59. 117 Ga. 191, 43 S.E. 419 (1903).

60. Id.

61. 23 ALR4th 674, 702.

62. Id.

63. 156 Ga.App. 209, 274 S.E.2d 653 (1980).

64. 253 N.Y. 73, 170 N.E. 498 (1930).

65. Id.

66. Id.

67. NY Const. Art. I, § 6, note 749.

68. Administrative Code of City of New York, § C26-388.0.

69. Bergen v. Morton Amusement Co., 95 Misc. 647, 159 N.Y.S. 935, aff'd, 178 App. Div. 400, 165 N.Y.S. 348 (1916).

EXCAVATIONS AND CONTAMINATION

Bryan P. Sweeney[1], M. ASCE
Joel S. Mooney[2], M. ASCE

ABSTRACT

This paper presents a discussion of environmental issues related to projects that require excavations on contaminated sites. Regulatory background information concerning contaminated soil and groundwater handling and disposal are reviewed. Environmental site characterization and consideration of contamination during project planning, design, and preparation of contract documents are discussed. Several case histories are presented to illustrate various ways recent projects were successfully completed on contaminated sites. As a basis for these discussions, the current regulatory setting and practices in the Boston, Massachusetts area are considered.

INTRODUCTION

This paper focuses on the critical issues related to successfully completing projects with excavations on contaminated sites. These issues include understanding the regulatory setting, characterizing the site, and considering the environmental concerns in the planning, design, and construction phases. Although federal regulations are consistent in the United States, the state, local, and municipal regulations typically vary at different locations. The major points of this paper are illustrated with respect to applicable environmental regulations in the Boston, Massachusetts area.

---

[1]Senior Engineer, P.E., Ph.D., Haley & Aldrich, Inc., 58 Charles Street, Cambridge, Massachusetts 02141

[2]Senior Engineer, P.E., Haley & Aldrich, Inc., 58 Charles Street, Cambridge, Massachusetts 02141

EXCAVATION AND CONTAMINATION

One primary environmental issue addressed within is the regulatory aspect of handling and disposing of soil and groundwater removed from excavations. The disposal options for excavated soil include landfilling, reuse on-site, and in some instances, batch processing petroleum contaminated soils at a recycling plant. Similarly, contaminated groundwater may be treated on-site and discharged, or collected and transported to a disposal facility. Alternatively, it may be more cost effective to remediate contaminated soil and/or groundwater; however, it is not within the scope of this paper to present and discuss remediation alternatives.

Successfully completing excavations in contaminated areas can be facilitated by:

o Understanding the regulatory setting.

o Performing environmental site characterization to determine if contamination exists on-site and characterizing the type, extent, and level of contamination prior to construction.

o Evaluating the project design relative to the potential impacts of contamination.

o Addressing the environmental issues in the contract between the Owner and Contractor (e.g., provisions in the contract documents) prior to construction.

o Scheduling and properly conducting the work during construction, considering the environmental issues.

These items are discussed below and, subsequently, several case history examples are presented.

REGULATORY SETTING

This section discusses federal, state, and local regulations related to handling and disposing of soil and groundwater removed from excavations. Some of the applicable regulations require that permits (e.g., dewatering permits) be obtained and/or certain investigations and reporting be conducted prior to performing the work. The controlling regulations and required permits are numerous, complex, and often site specific, depending in part on:

o The type and level of contamination.
o The proposed disposal approach.
o The status of the site relative to applicable federal and state regulations.

Applicable laws and regulations are discussed below followed by discussions of specific regulations related to soil disposal and groundwater disposal.

A. Regulations

In the Boston area, sites with contamination are subject to the Massachusetts General Laws (MGL) Chapter (c.) 21E and respective regulations, the Massachusetts Contingency Plan (MCP) promulgated at 310 Code of Massachusetts Regulations (CMR) 40.000 (Reference 1), and MGL c. 21C and the Massachusetts Hazardous Waste Management Regulations promulgated at 310 CMR 30.000 (Reference 2). The Massachusetts Department of Environmental Protection (DEP) has jurisdiction over both sets of regulations.

MGL c. 21E and the MCP primarily apply to sites which have had a release or present a threat of a release of oil or hazardous material. If the site is in the MCP process prior to construction the DEP will oversee the environmental-related project issues. However, a localized spill with a definable source may be immediately remediated outside of the comprehensive and sometimes lengthy phased MCP process under the Emergency Spill Section of the MCP (Reference 3). MGL c. 21E is similar to the federal Comprehensive Environmental Response, Compensation, and Liability Act (CERCLA, Reference 4) which is administered by the Environmental Protection Agency (EPA), and the National Contingency Plan (NCP) promulgated at 40 Code of Federal Regulations (CFR) 300 (Reference 5).

MGL c. 21C and the regulations at 310 CMR 30.000 (Reference 2) primarily have jurisdiction over active hazardous waste generators, transporters, and treatment, storage, and disposal facilities. These regulations establish standards for hazardous waste determination, labelling, storing, treatment, and disposal. With the exception of the corrective action and land disposal restrictions programs, MGL c. 21C and regulations are equivalent to, and in some instances more stringent than, the federal Resource Conservation and Recovery Act (RCRA), and RCRA regulations promulgated at 40 CFR 260-280 (Reference 6).

In addition to the federal and state laws and regulations pertaining to hazardous waste management and waste site remediation, there are federal and state regulations applicable to the discharge of water from excavations. Regulatory agencies

EXCAVATION AND CONTAMINATION 29

requiring permits for discharges of groundwater to a surface water body, sewer system, or the ground are the federal EPA, DEP, Massachusetts Water Resources Authority (MWRA), and/or the Boston Water and Sewer Commission.

B. **Specific Regulations Related to Soil Handling and Disposal**

Typically, the soil removed from excavations can be classified into the following four general categories:

o Hazardous Waste
o Oil-Contaminated/Hazardous Materials
o Regulated Soil
o Unregulated (Clean) Soil

State regulations and criteria associated with these categories, with the exception of "hazardous waste", are implemented by the DEP Bureau of Waste Site Cleanup (BWSC). Soil contaminated with hazardous waste is regulated by the DEP under MGL c. 21C and 310 CMR 30.00 (Reference 2).

The BWSC has numerous policies regarding the treatment and disposal of soil, which depend in part, on the type, extent, and source of contamination (Reference 7). In addition, they work closely with the DEP Solid Waste Division that licenses landfills and recycling facilities within the state. Excavated soil must also meet criteria of the landfill or disposal facility.

Four general categories and disposal criteria for excavated soil are discussed below. Figure 1 presents a conceptual flow chart that indicates various disposal options associated with these categories. Typically, the excavated soil is categorized in the field based on criteria in the Excavated Soil and Dewatering Management Plan (ESDMP). This plan is discussed in a subsequent section.

1. **Hazardous Waste**

Hazardous waste is defined under RCRA and MGL c. 21C regulations. A waste is identified as a RCRA hazardous waste based on any of the following four characteristics: ignitability (310 CMR 30.122 (Reference 2), 40 CFR 261.21 (Reference 6)), corrosivity (310 CMR 30.123 (Reference 2), 40 CFR 261.22 (Reference 6)), reactivity (310 CMR 30.124

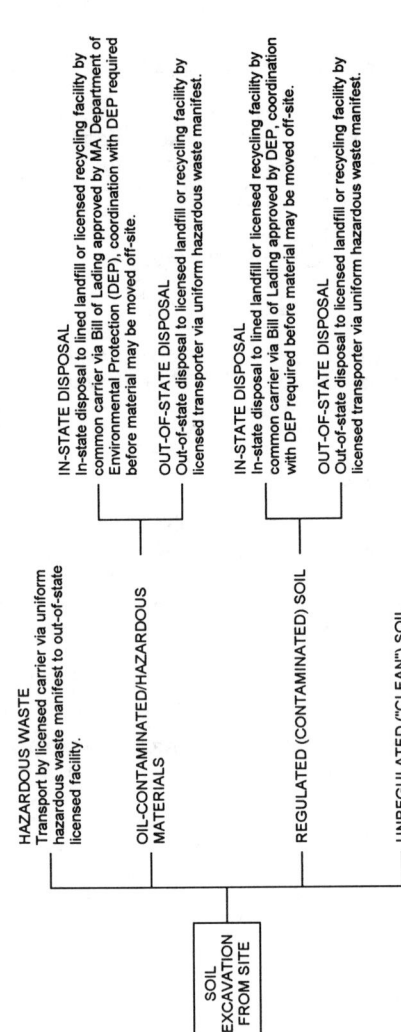

Figure 1. Conceptual Paths for Soil Disposal

NOTE: Assumes appropriate regulatory compliance on behalf of generator prior to disposal.

(Reference 2), 40 CFR 261.23 (Reference 6)), or toxicity (310 CMR 30.125B (Reference 2) and 40 CFR 261.24 (Reference 6)).

The toxicity characteristic is determined based on testing the leachate generated from the Toxic Characteristic Leaching Procedure (TCLP). RCRA hazardous wastes are listed at 40 CFR 261.31 through 261.33 (Reference 6), or 310 CMR 30.120 through 30.136 (Reference 2). A waste is also defined as a RCRA hazardous waste if it appears on any of the lists in 310 CMR 30.130 through 30.136 (Reference 2). Soils that are mixed with or derived from hazardous waste are classified as hazardous waste and are regulated per 310 CMR 30.00 (Reference 2) and the equivalent RCRA regulations at 40 CFR 260-280 (Reference 6).

Soil classified as RCRA hazardous waste must be disposed of at a licensed hazardous waste disposal facility and transported under a Uniform Hazardous Waste Manifest (UHWM). In Massachusetts, soil that is classified as RCRA hazardous waste must be disposed of out-of-state since no licensed hazardous waste disposal facilities currently exist in state.

2. **Oil-Contaminated/Hazardous Materials**

    Oil-contaminated/hazardous materials, as defined by MGL c. 21E, result from specific releases or spills of oil or liquid chemicals and/or wastes. Soils are classified as oil contaminated or hazardous materials based on the level and types of contaminants established by the DEP, at 310 CMR 40.00 (Reference 1).

    These materials also require out-of-state disposal and a UHWM for transport or disposal at an in-state treatment facility. Excavated soil that meets the definition of an oil-contaminated or hazardous material, and a RCRA hazardous waste, is governed by both sets of regulations.

3. **Regulated and Unregulated Soils**

    Regulated and unregulated soils are classified based on criteria established by the DEP (including References 2, 7, 8). Unregulated soils are considered clean or equivalent to background conditions and may be reused as backfill.

Regulated soils may have low levels of metals or other contaminants that cannot be attributed to a specific release. These soils may be disposed of in-state or out-of-state. A Bill of Lading must be obtained from the BWSC for transportation of regulated soils to an in-state facility. Out-of-state transport to a disposal facility must be performed by a licensed transporter under a UHWM.

C. Regulations Related to Discharge of Groundwater

This section discusses regulatory and permit issues applicable to groundwater and also precipitation pumped from excavations. Applicable regulations and permits which govern groundwater discharge depend upon the proposed discharge alternative.

The discharge of contaminated water is generally not acceptable until the levels of contaminants are reduced below allowable maximum contaminant levels indicated or referenced in the discharge permit. Therefore, on-site groundwater treatment is typically necessary prior to groundwater discharge from a contaminated site, unless the groundwater is disposed of at an off-site facility. Discharge permits associated with treated groundwater generally require that a suite of chemical tests be conducted on a periodic basis during the discharge period, with more tests typically performed during the initial discharge phase.

- Discharge directly to a surface water body.
- Discharge into the municipal combined storm drain and sewer system that routes the discharge to:
    a. A surface water body, or
    b. The MWRA treatment plant.
- Off-site disposal.
- On-site recharge.

These discharge alternatives are discussed below and a conceptual flow chart that indicates the various options is presented in Figure 2.

1. Discharge into Surface Water Bodies

Discharges to surface water bodies require National Pollutant Discharge Elimination System (NPDES) permits, which are issued jointly by the EPA and DEP. Applicable regulations include the federal NPDES regulations promulgated at 40 CFR Part 122 (Reference 9) and the Massachusetts DEP Division of Water Pollution Control regulations at

EXCAVATION AND CONTAMINATION

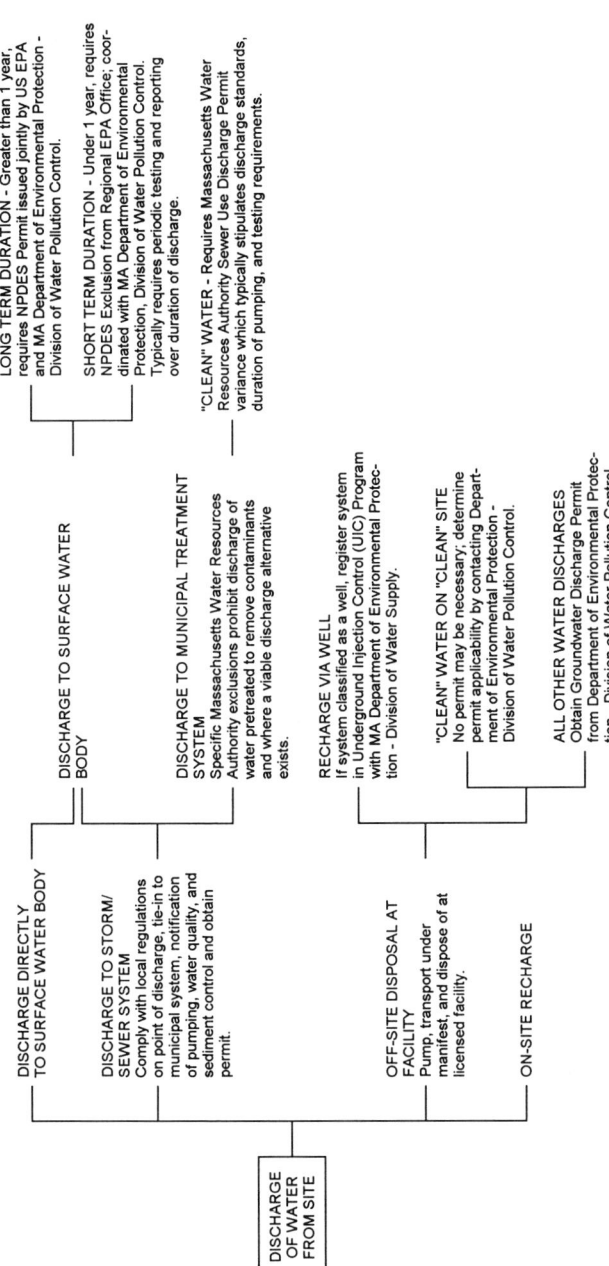

Figure 2. Conceptual Paths for Water Discharge

NOTE: Assumes water (if contaminated) has been treated prior to discharge to remove chemical constituents to concentrations below permissible levels, and appropriate regulatory compliance on behalf of generator prior to disposal.

314 CMR 3.00 (Reference 10). The NPDES permits are initially reviewed by the EPA and subsequently by the DEP, unless an exclusion is obtained.

Applications for NPDES permit exclusions are often submitted for temporary discharges associated with short-term excavations. An NPDES exclusion for temporary discharges does not need DEP approval and may be obtained from the EPA within a relatively short time period.

Approved permit exclusions typically specify sampling and analysis requirements and standards the discharge must meet. The effluent standards are often set at drinking water standards and discharges to surface waters must meet the State Surface Water Quality Standards at 314 CMR 4.00 (Reference 11). The NPDES permitting process may extend over one year for the evaluation of long-term discharges.

2. Discharge into Municipal System

   Discharges into the storm water and/or sewer collection system are under the immediate jurisdiction of the Boston Water and Sewer Commission. The pipes in this system generally route the discharge into a surface water body or to the MWRA treatment plant. Discharges to surface water bodies are discussed above.

   The Boston Water and Sewer Commission has promulgated regulations that include prior written approval for discharge to their infrastructure and include standards the discharge must meet. Sediment loading is a major concern and usually some form of sediment removal (e.g., sedimentation tanks) is necessary prior to discharge.

   A permit application must be submitted to and approved by the MWRA prior to discharge. Permits for discharge to their Deer Island facility are difficult to obtain because of the current high volume demands at the plant. Furthermore, the MWRA has made a policy decision to prohibit any groundwater that requires treatment prior to discharge (Reference 12). In addition, the MWRA will prohibit discharges to their system if other discharge options exist.

EXCAVATION AND CONTAMINATION   35

3. **Off-Site Disposal**

   Contaminated groundwater may be disposed of
   off-site at a licensed facility with transport via
   a licensed hauler and a UHWM.

4. **On-Site Recharge**

   Recharge of water on-site is regulated by the
   DEP. In general, water recharged to the ground
   must meet Groundwater Quality Standards of 314 CMR
   6.00 (Reference 13). Furthermore, the manner and
   configuration which the water is recharged (e.g.,
   wells or horizontal drains) may dictate additional
   requirements such as registration of the system
   under the Underground Injection Control Program
   with the DEP Division of Water Supply. Also, the
   discharge system may be subject to the DEP
   Division of Water Pollution Control policies and
   regulations.

   If the groundwater is determined to be
   contaminated and treatment is required prior to
   discharge, a groundwater discharge permit must be
   obtained from the DWPC pursuant to 314 CMR 5.00
   (Reference 14). This permitting process also
   requires a considerable time period. Discharge
   standards are usually set at drinking water
   standards.

ENVIRONMENTAL SITE CHARACTERIZATION

The issue of on-site contamination should be evaluated
in the initial project phases, and not only when
required by regulations. A common pitfall in the
project planning phase is to "bury your head in the
sand", i.e. performing little to no environmental
evaluation and hoping environmental issues will not
impact the project. Generally, this is not a
satisfactory approach for several reasons, including
worker health and safety during construction, cost
impacts, schedule delays, and potential regulatory
fines. Environmental site characterization (ESC) should
be priority item conducted early in the project at a
level that is appropriate for the site and project.

The ESC will develop and evolve as the findings of the
site investigation are gathered. Initial steps of the
ESC typically include the following sequential steps:

o   Review of the site history, and also site history of
    adjacent parcels, using available information such as

insurance maps, tax maps, town records of site use, fire department information on buried tanks, etc.

o   Observations of current site usage and conditions on and adjacent to the site.

o   Review of available subsurface conditions in the area using published geologic information, other records, and recent explorations. Environmental and geotechnical subsurface investigations can oftentimes be combined to reduce the overall costs associated with two separate programs.

o   If contamination is suspected, conduct preliminary field and laboratory investigations to assess the type and extent of the contamination.

The level of subsequent evaluations will typically vary for each site based on the information gathered and regulatory setting. If contamination is present, subsequent steps of the ESC should be consistent with applicable regulations and gather information to estimate the concentrations and volume of contaminated soil and groundwater, including the lateral migration and vertical extent.

The ESC should be performed to provide sufficient information to assess the potential for contaminated soil and/or groundwater and determine the need for regulatory involvement/interaction. In some instances, the construction schedule and excavation may be significantly influenced by the regulatory process.

PROJECT DESIGN

Environmental issues associated with contaminated soil and groundwater should be considered by the project team in the design phase. Early assessment of environmental concerns will help to allot the necessary costs, adjust the project schedule, and include additional work associated with the disposal of the contaminated materials.

The project schedule and budget should include allowances for the time and costs associated with regulatory compliance, including items related to filing, review, preparing permit applications, and interaction with the applicable agencies and, if necessary, the public. Contingencies for additional environmental investigation efforts may also need to be included, depending on the confidence level in the ESC, and anticipated regulatory requirements. In some instances, it is necessary to educate the Owner on the importance of addressing the environmental issues and performing the necessary investigations.

Adequately addressing these issues in the design phase will potentially reduce construction costs. Also, consideration of environmental issues at this stage and their inclusion in the general conditions and contract documents should reduce change orders and delays during construction. The design team can also estimate and bid the work to evaluate costs associated with the environmental issues.

To reduce the impacts of contamination on construction schedule, sequence, and cost, excavations in contaminated areas should be reviewed and alternative design and construction schemes should be evaluated. Alternative schemes may include:

o   The deletion or reduction of below-grade space to reduce the volume of contamination encountered in the excavation.

o   Utilize foundation types that reduce the need for excavations, such as pile foundations.

o   Design and install relatively impermeable excavation support walls (e.g. slurry walls, or cut-off walls) to reduce the volume of contaminated groundwater pumped within the excavation.

o   Relocate the below-grade structure areas to portions of the site where the environmental considerations would be reduced.

o   Design the below-grade structures for full hydrostatic pressures to avoid long-term groundwater pumping and change in hydraulic gradients around the structure.

o   Design and install a groundwater recharge system to temporarily alter the hydraulic gradient to reduce the amount of contaminated groundwater that could migrate from off-site sources.

CONTRACT CONSIDERATIONS

This section discusses environmental-related contract issues and documents that should be included in the agreement between the Owner and Contractor. Adequately addressing environmental issues in the contract documents, including pay items for environmental-related items and incorporating this into the overall project, will typically reduce the costs associated with contract extras and change condition claims, and also will reduce construction delays.

A. Contract Issues

It is difficult, if not impossible, to prepare a contract between the Owner and Contractor that will completely cover all the work and payment items related to excavations, particularly for projects with environmental concerns. However, contracts that are flexible include compensation and/or allowances for environmental work and share risk between the Owner and Contractor, will generally result in lower construction costs.

These types of contracts may reduce cost contingencies carried by the Contractor that account for unknown or unforeseen conditions. Otherwise, prospective bidders may include larger environmental cost contingencies associated with handling and disposal of soil and groundwater that will exceed the project budget, and may or may not be sufficient. In contrast, the low bidder may not have included any environmental-related costs and claims for extra costs will surely follow.

Examples of contract flexibility include environmental-related payment items in unit price contracts. In lump sum contracts, well defined work scopes with allowances for items that can not be accurately defined prior to construction would provide this flexibility.

For example, the earth support, excavation, and dewatering aspects of the project would be bid lump sum for the baseline scheme of specific configuration. Additional work for excavation support and excavation would be bid at unit prices. Dewatering costs would assume uncontaminated water with a unit price included for treating and/or handling contaminated water. Off-site disposal of contaminated soil and groundwater would be paid for by unit prices or from other established allowances in the contract. Different unit price items may be included for various disposal options which depend on the type and level of contamination, e.g. out-of-state disposal, disposal at a recycling facility, or disposal at a in-state lined landfill.

As indicated in this example, the Contractor is typically responsible for the disposal of

EXCAVATION AND CONTAMINATION 39

contaminated materials because the Contractor handles these materials. Manifests and Bills of Lading are typically obtained by the Contractor.

B. Contract Documents

The remainder of this section discusses environmental considerations that should be addressed in the project plans and specifications. These documents should include or reference the following:

o The information and data gathered in the Environmental Site Characterization and documents submitted to regulatory agencies.

o An Excavated Soil and Dewatering Management Plan (ESDMP) that defines excavation, transport, and disposal requirements of materials in addition to health and safety considerations.

o Project specification sections that are prepared with consideration to the environmental issues.

o In some cases, permits obtained by the Owner and information on those required by regulatory agencies.

The ESC information should be included to inform the Contractor of the type, concentration levels, and location of observed and anticipated contamination. This information is necessary for the Contractor to schedule various construction activities associated with contaminated excavations, obtain the necessary permits, if any, and to determine appropriate means and methods to perform the work safely.

1. Excavated Soil and Dewatering Management Plan

A site specific ESDMP should be prepared to define requirements for soil excavation and dewatering procedures. The procedures and criteria established in this document must be consistent with federal, state, and local regulations. In Massachusetts, this plan is submitted to the DEP for review and comment prior to construction. This submittal is necessary to obtain a Bill of Lading from the BWSC to permit transportation of contaminated soil and groundwater for in-state disposal.

The ESDMP is site specific and addresses the following six principal areas of concern:

- o Field Soil Monitoring Requirements (during excavation)
- o Soil Handling, Storage and Disposal
- o Transportation of Materials
- o Health and Safety
- o Disposal of Surface Water and Groundwater
- o Contingencies

The soil and groundwater monitoring requirements depend upon the type and level of contamination. The monitoring program may include field screening, visual, and olfactory identification and, if necessary, confirmatory laboratory testing. Typically, this program is performed by a representative of the Owner.

Soil handling, storage, and disposal options are provided for the various soil categories in the ESDMP that are based, in part, on the type and level of contamination. Categories of soil anticipated to be encountered are developed from ESC data. The soil encountered in the excavation is then categorized based on the monitoring results, and each category may be handled and transported according to different pre-approved procedures.

Temporary soil stockpiles may be necessary on site for soils that must be removed from the excavation, but require additional testing to determine the appropriate disposal destination. These stockpiles are typically placed on and covered with polyethylene sheeting.

Health and safety criteria (Reference 15) govern human interaction with hazardous or toxic materials and dictate monitoring of various conditions (e.g. volatile organic compounds in the air) to protect on-site workers and the public. Contingencies are necessary in case potentially hazardous conditions or conditions not anticipated (based on the ESC) become evident. If such conditions are encountered and contingencies have not been developed, the work may be suspended until the environmental issues are evaluated and health risks addressed.

2. Project Specifications

After the ESDMP is reviewed and approved by the DEP, the Earthwork, Construction Dewatering, Health and Safety, and other specification sections can be prepared to define the requirements under which the work is to be performed. These sections should also include payment provisions for handling, trucking, and disposal of contaminated soil and groundwater.

The Earthwork specification should include requirements for on-site handling and stockpiling, and as necessary, field screening, and possibly chemical testing. The Health and Safety section should require the Contractor to prepare a plan to monitor and protect the health and safety of the workers. This plan is discussed in the construction section of this paper.

The Construction Dewatering section should include requirements for groundwater screening and testing. These requirements must be in accordance with applicable permits, particularly if the permit was obtained by the Owner. Groundwater discharge permits are sometimes obtained by the Owner to reduce delays associated with permit application and review by regulatory agencies. Permit applications submitted after the bid is awarded may delay the start-up of construction.

CONSTRUCTION

As previously emphasized, environmental issues should be evaluated during design and prior to construction to reduce the potential for encountering unforeseen contamination during excavation. The project schedule should consider the necessary time associated with environmental issues in order to properly sequence the work and evaluate the project's critical path. These items should be planned considering the potential for delays.

Prior to construction, the Contractor should be required to submit a Health and Safety Plan for review by appropriate members of the project team. This submittal indicates the Contractor's program to protect the health and safety of all workers associated with contaminated soil and groundwater.

Excavations in contaminated areas should be planned to eliminate the possibility of cross contamination of "clean" materials during excavation, stockpiling, and trucking. For example, it may be prudent to entirely excavate contaminated soil layers or pump groundwater within the excavation limits before proceeding deeper in local areas and consequently contaminating the underlying soils and/or groundwater. To minimize this, excavation may often begin at areas of greatest contamination and work towards areas of lesser contamination.

The disposal of surface water and groundwater should be conducted in compliance with applicable regulations and permits. The permit will specify sampling and testing for certain chemical constituents on a scheduled basis.

Definitive measures should be taken to control surface water that results from precipitation (rain or snow) during excavation to minimize water that requires handling and possibly treatment. Also, off-site contamination may migrate on-site due to the seepage gradients created by an excavation conducted below the groundwater level. Lateral support systems, dewatering methods, and construction sequencing should be reviewed in the design phase, and the reviewed system(s) should be implemented during construction to reduce the pumping of contaminated groundwater.

Field monitoring during excavation is necessary for several reasons including health and safety purposes, regulatory and permit compliance, to effectively categorize soil for disposal (thereby reducing stockpiling), and to determine a suitable disposal alternative.

CASE HISTORY EXAMPLES

The following examples are based on recent experience in the greater Boston area where environmental issues influenced project planning, design, and/or construction.

o In order to avoid handling and disposal of contaminated soil, a below-grade mechanical area originally designed with conventional spread footings was reduced in depth, made larger in plan area, and redesigned with a mat foundation. The mat foundation was necessary in the less competent bearing soils. Although a premium was paid for the mat, the overall project cost was less than disposal of the contaminated soil and groundwater associated with the deeper excavations necessary for less expensive foundation support.

o At another site, short piles (less than 10 ft. long) were driven through a near surface contaminated soil layer to achieve bearing in competent soils below. Piles represented a least cost alternative when compared to costs of excavation, handling, characterization for disposal, disposal, interaction with regulatory agencies, and potential for construction delays.

o Early characterization of the groundwater quality and its impacts on construction dewatering allowed time to specify an on-site remediation process in the contract documents and to obtain necessary permits from regulatory agencies. The proposed treatment and an effluent monitoring program was discussed with regulatory agencies and implemented. This facilitated groundwater discharge into the storm drain system. While regulatory interaction and compliance and remediation were premiums, the project schedule was not adversely effected.

o After the concentration and extent of the contamination was determined at a building site, an early site remediation contract was let. Excavation and disposal began at the area of greatest contaminant concentration, with water being pumped out of the excavation by a vacuum truck. The remedial excavation process radiated outward from this area to remove contaminated materials. To restore the excavated area to a grade compatible with later building foundations, the over-excavated area was backfilled with lean concrete.

o To minimize contaminated groundwater flow into the excavation, lateral earth support was comprised of interlocking steel sheet piling with internal bracing below the water table. In changing the bracing method from external (tiebacks) to internal bracing, contaminated water was not generated during tieback installation. Internal bracing also minimized penetrations through the earth support system and the sheet pile wall acted as an effective cutoff. Penetrations in the sheet pile wall were sealed by grouting.

SUMMARY AND CONCLUSIONS

Environmental issues should be evaluated early in the project to reduce project costs, construction delays, change orders, and extra work associated with the disposal of contaminated soil and/or groundwater.

The major steps involved in reducing the impacts of environmental issues on a project relative to the disposal of soil and groundwater consist of the following:

- Understand all applicable regulations and interact with regulatory agencies early in the project in an attempt to anticipate their requirements.

- Comply with applicable regulations.

- Perform an ESC to characterize the site. The ESC may be performed in phases that are coordinated with the project design and/or regulatory requirements.

- Consider design alternatives or modify construction methods to reduce or minimize the contamination encountered or generated.

- Prepare a contract that communicates the scope of the environmental issues and the expectations of the design team to the Contractor. Adequate technical and payment provisions should be provided for flexibility to accommodate changes in the field.

- Schedule and conduct the work considering the potential effects of contamination.

Projects that are planned with consideration of environmental issues will have lower overall costs and less delays and reduced environmental liabilities for the Owner and Contractor than projects where the project team neglects to evaluate environmental issues or "buries their heads in the sand".

ACKNOWLEDGEMENTS

The authors gratefully acknowledge the assistance of Mr. Mark X. Haley and the beneficial review comments provided by Ms. Nancy Prominski in the preparation of this paper.

REFERENCES

1. 310 CMR 40.00, Massachusetts Contingency Plan.
2. 310 CMR 30.00, Hazardous Waste Regulations.
3. "Guidance on Differentiating Disposal Sites from Spills," Policy No. WSC-89-002, DEP, BWSC.
4. Comprehensive Environmental Response, Compensation, and Liability Act (CERCLA).
5. 40 CFR 300, National Contingency Plan.
6. 40 CFR 260-280, Subchapter I - Solid Wastes (Continued).
7. "Update on Bureau of Waste Site Cleanup Policies," Massachusetts Department of Environmental Protection, August 16, 1991.
8. 310 CMR 19.00, Solid Waste Management Regulations.
9. 40 CFR Part 122, Federal National Pollutant Discharge Elimination System (NPDES) Permit Regulations.
10. 314 CMR 3.00, Massachusetts Surface Water Discharge Permit Rules.
11. 314 CMR 4.00, Massachusetts Surface Water Quality Standards.
12. 360 CMR 10.00, Sewer Use Rules and Regulations.
13. 314 CMR 6.00, Massachusetts Groundwater Quality Standards.
14. 314 CMR 5.00, Groundwater Discharge Permit Program.
15. 29 CFR 1910.120, 29 CFR 1910.1000-1101, 29 CFR 1910.134, Occupational Safety and Health Standards.

OPPORTUNITIES AND CONSTRAINTS FOR THE
INNOVATIVE GEOTECHNICAL CONTRACTOR

Peter J. Nicholson[*], M.ASCE, and
Donald A. Bruce[+], M.ASCE

ABSTRACT

The American specialty geotechnical construction community has historically been followers rather than leaders. This is due to the nature of past construction demands, and to current litigious and confrontational operating conditions. Today the demand for innovation and sophistication is great, and likely to grow, but the industry is facing major problems associated with low profitability. This paper reviews generic options for survival and growth for innovative contractors. From an internal viewpoint, companies can benefit from the principles of Total Quality Management, improved response to customers, and appropriately structured innovation. From an overall industry viewpoint, all will benefit from alternative bidding practices and dispute resolution, and the new process of Partnering. The principles of each of these options are summarized and illustrated with reference to recent projects.

1. INTRODUCTION

When considering innovative processes in specialty geotechnical contracting, one observation is common, namely the virtual absence of U.S. origins (Nicholson, 1986; Bruce, 1988, 1992a, 1992b). One notable exception is in the field of ground treatment, where compaction grouting, referred to by Baker et al. (1982) as a

---

[*] Chief Executive Officer, and [+] Technical Director, both of Nicholson Construction of America, P.O. Box 308, Bridgeville, PA 15017

"uniquely American process", is now being exported following its development in the early 1950's (Warner, 1982, 1992).

In contrast, most of our techniques, from ground anchors to pinpiles, and from diaphragm walls to soil mixing and jet grouting have been introduced from Europe and Japan. Once these techniques were introduced, however, - largely through the perseverance of sponsoring contractors - they acquired their own distinctly American flavor; maturing in response to demands on scale, speed and economics (e.g., Bruce and Nicholson, 1988). The fact remains, nevertheless, that the U.S. is not renowned internationally as an innovative geotechnical community, despite the adaptability, effort, and resources which are available here.

The reasons for this are complex and deep-rooted, but broadly may be delineated as follows:

- Lack of Necessity. Traditionally we have not had to conceive novel and highly sophisticated technologies to accomplish our major national building tasks, such as the construction of the interstate highways or the railways, or our network of river controls. The country is large, and the population density is about 100 times lower than in industrialized Europe or the Far East. We have often had the luxury of selecting from a number of suitable sites, and this has encouraged the solution of simply avoiding difficult conditions. Now, however, the emphasis has changed within our business to urban and industrial redevelopment, infrastructure upgrading and improvement, and enhanced transportation facilities. One consequence is that we have a growing market for techniques developed abroad during reconstruction after major wars, or in response to rapidly growing population centers. Geotechnical problems must be solved where they occur: relocation is usually not an option.

- Contractural and Legal Processes. Most projects are still awarded to contractors who submit the lowest cost estimate for the work they believe they have to carry out, but which may not be the same work foreseen by the owner. Calculation of a low bid carries no guarantee on the ability to perform satisfactory quality work and yet this "low bidder" paradigm of the consumer society persists. The disillusionment and confrontation arising from this approach have been reflected in a substantial growth of construction disputes. The Arbitration Times (Winter 1990/91) reports that claims in construction cases increased by 16% from 1988 to 1989 while mediation cases rose 38% in the same period. Total cases submitted to mediation increased fourfold from 1987 to 1989.

According to <u>Business Week</u> (April 13, 1992), the U.S. has 307 lawyers per 100,000 population - three times more than in Britain, and 25 times the Japanese figure. The 1971 census of over 355,000 lawyers compares to over 750,000 in 1990 and a projected 1 million in the year 2000. Our typical contracting methods foster adversarial attitudes, and the reliance on lawyers to resolve these issues deters all but the most hardy and committed contractors. Quite simply, this litigious atmosphere is not compatible with the spirit of innovation, for the real risks far outweigh the potential rewards.

There is a growing need in this country for specialists with the skills necessary to solve a wide range of geotechnical problems. Judging from the scale and complexities of environmental remediation, these skills are equally valuable in that sector also. And yet, set against this demand, is the plain fact that the contracting industry is in disarray: some companies have gone out of business, many are having runs of "bad years", and most are not as profitable as they feel they should be.

The authors believe, nevertheless, that there are several options open to the innovative contractor who wishes to survive and grow. Some of these options have an internal focus, while others require bilateral cooperation. The former group include Total Quality Management (TQM) and continuing innovation. The latter group includes alternative bidding practices, Partnering, and dispute resolution.

The authors also believe that while most of our profession have heard many of these terms, few really appreciate the basic principles and the impact they can have on our industry. It is the purpose of this paper to provide an introduction to each of these concepts in turn, and to illustrate their applicability with respect to the authors' recent experiences. No credit is claimed for the development of the concepts themselves, as evidenced by the references and acknowledgements. However, the authors trust the reader will appreciate the originality of addressing this vital issue within a progressive, quality conscious and cooperative framework, as opposed to the confrontational tone common in essays such as "The Contractor's Viewpoint" or "The Consultant's Viewpoint".

2. TOTAL QUALITY MANAGEMENT (TQM)

According to proponents of TQM, a "revolution is brewing in American business - one as important to our times as the automobile and the steam engine were to theirs. This revolution is called Quality, and it is reshaping the way we think about everything we do." (Dobyns and Crawford-Mason, 1991). Cynics would say it is simply the current fad, and would in fact suggest that it is already obsolescent. They would argue that it is simply another transient phase in a tenuous path including assurance and control (quality cannot be inspected into a product) and quality circles (collapsed due to the lack of commitment from top management).

However, there are two key issues which cannot be denied. Firstly, there is no doubt that there is a growing awareness of, and real need for, quality in our industry. The quality of the end product reflects the quality of the processes which interact to produce it: the key to each process is people - their attitudes, their commitment, their training, and their management. Secondly, as demonstrated repeatedly by the gurus of the movement - W. Edwards Deming, Philip B. Crosby, Armand V. Feigenbaum, and Joseph M. Juran - and by the success of our "enlightened" companies such as Motorola and Xerox, as well as most of Japanese industry, there is equally no doubt that attention to quality reduces costs by increasing productivity through minimizing rework.

In our industry, this means doing things correctly the first time, at every stage in each process. Designs must be technically correct but practically constructable; bid documents and specifications must be complete, and clear in defining what is expected of the bidder; the bidder must be equally clear in his response and have no "hidden agendas" or conditions on rock bottom prices left trailing as hooks for future claims; the contractor and the site supervision must build what is required where it is required and when it is required; the owner must pay, on time, the amounts actually due, not those manipulatively recalculated to enhance his cash flow. Within each of these groups of parties, quality processes have to be encouraged, and full cooperation and understanding between design, estimating, engineering, construction and administration departments are essential.

Several educational facilities and institutes actively offer training in the concepts and details of TQM. One such short course is offered by Fails Management Institute (FMI), which defines TQM as " a systematic process for continuous improvement throughout the organization."

These words have been precisely chosen, and those underlined (by the authors) each have special significance. FMI summarizes that there are six basics on which the TQM process is built:

1. Every job activity is a process that includes inputs and outputs, suppliers and customers. A corollary is that everyone in the organization should regard himself/herself as everyone else's client and customer. Also inherent is the fact that we have internal _and_ external clients and customers, and that the quality of response we each should give external contacts should be mirrored in our dealings _within_ our own organization.

2. Quality is compliance with the customer's requirements. The requirements are error-free work. In certain ways, compliance is a minimum standard, but it must always be realized that quality is not synonymous with perfection.

3. The method for achieving error-free work is prevention. This reverts to the earlier discussion, and can be studied in four steps:

   - examine the job activity as presently conducted
   - determine the variance from an error-free performance
   - establish a new prevention process
   - measure (and display) the results of the job activity.

4. The cost of quality is measurable, and equals the cost of errors plus the cost of prevention. The cost of errors includes delays, lost customers, accidents, rework, idle time, and litigation. The cost of prevention is in training, planning, supervision and testing. Figure 1 is an excellent representation of the financial impact of reducing the cost of errors: the data are in fact in line with those truly recorded in the construction industry. Basically, the reduction in the cost of errors translates almost wholly to the bottom line since the saving is proportionally much larger than the additional expenditure needed to improve quality.

5. Quality, productivity, and safety are inseparable. A safe act may not be a quality act because it may not be a productive act. A productive act may not be a quality act because it may not be a safe act. However, a quality act will include safety and productivity.

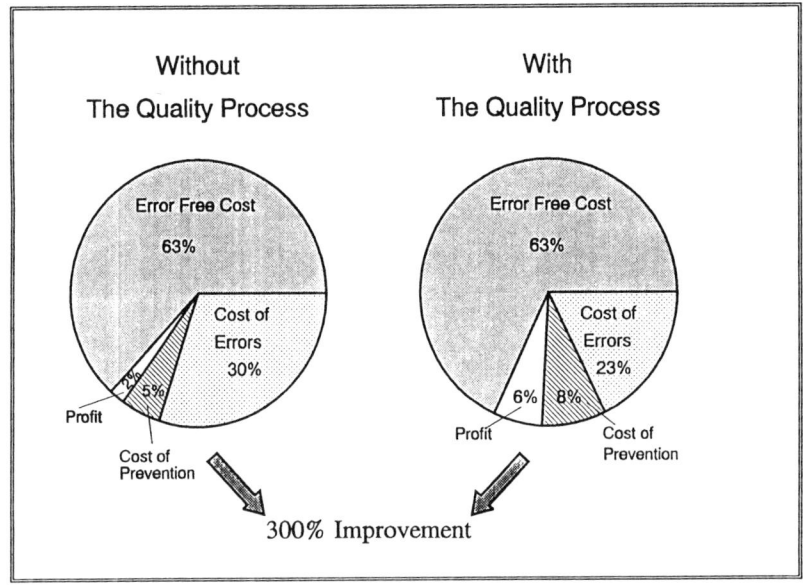

Figure 1. Illustrating how a quality plan acts to reduce the "Cost of Errors" and so increases profitability (courtesy of Fails Management Institute)

6. The keys to quality are commitment and teamwork - again the human factor. Dobyns and Crawford-Mason (1991) are especially interesting in this regard, and conclude "how well you educate, train and treat people in your society becomes more important than the coal you dig, trees you fell or rivers you dam." Equally, the MIT Commission on Industrial Productivity says, "There seems to be a systematic undervaluation in the United States of how much difference it can make when people are well educated and their skills are continuously developed and challenged."

Each one of these six basics has a clear and vital message to us in industry, and especially to those of us caught in the rythymns of entrepreneurial geotechnology. So fundamental are these basics that the term TQM is arguably not truly reflective of the subjects it covers. Perhaps ACS (Applied Common Sense) would be more apposite.

Some of these basics are for individual parties to address, and education and productivity are prime examples. However, most involve the active cooperation and interplay of all the contracted parties. These particular areas of common effort are discussed in the following sections, from which it becomes clear that one of the most important opportunities for survival we have is, simply, our mutual common sense and responsibility.

A final point relating to the principles of TQM is the issue of serving customers most efficiently. Companies must establish who their customers really are and what exactly they wish to buy. The key to this process is clearly information on the industry - often more important than the components of the industry themselves. For example, Dun and Bradstreet sold the Official Airlines guide for three times what TWA paid for Ozark Airlines and what Trump paid for the Eastern Shuttle.

Such information in our industry is often most easily gleaned or shared at the conferences, committees or publications of learned and professional societies. The innovative contractor will therefore find active participation in these societies rewarding from many viewpoints.

3. INNOVATION: CONSTRAINTS, BENEFITS AND REWARDS

For innovation to flourish, there must be a need and a reward for those who take the risk of funding and implementing the technology. One of the biggest constraints on this is our most commonly used delivery system for a construction project, the Design-Bid-Build System. The steps are well defined:

o **Design:** The owner, or a consultant selected by the owner, designs the project.

o **Bid:** A suitable bidding period is established, usually 4-6 weeks, and any contractor who can secure suitable bonding bids the work.

o **Build:** The low bidder is then selected and the project is built in accordance with the plans and specifications of the owner and consultant.

Along the way there may be some opportunities to innovate - the engineer or constructor can propose value engineering - but these opportunities are limited.

This present low-bidder system, used for the overwhelming majority of public works construction in the United States has led to other problems in the industry, in addition to the lack of innovation. Disputes, formal arbitration and lawsuits are becoming an everyday occurrence for many contractors. Low profit margins, given the inherent risks of construction, have led to the failure of many companies.

Another factor is that the constructon industry is fragmented to the extent that 1.3 million firms split an annual volume of $450 billion (U.S. Census Bureau). The largest companies still have a miniscule share of the overall market. Based on the reporting of the "Top 400 Contractors", the largest five contractors control less than 10% of the overall domestic market (ENR, 1991). This fragmentation of the industry and low profitability have meant that there is rarely money for research and development. Without research, innovation relies on happenstance or the inventiveness of individuals or small teams of people ("skunk works" in the words of Peters, 1984). As the industry consolidates, it is arguable that we will see larger contractors control more of the market, as is now true in Europe and Japan.

Bonding and financial requirements for the larger public works contractors are becoming more rigorous and this has led to more companies being operated like businesses: many constructors and consultants are now run by financial managers and MBA's. There are still many managed by engineers or tradespeople turned owners, but in the larger, more successful businesses, these entrepreneurs have learned to adapt and be more business oriented. The banks and bonding companies are asking to see detailed business plans and projections of activity and profitability. Financial reporting is expected on a monthly, or at least quarterly, basis, and is assumed to be accurate and without too many surprises - always a dangerous assumption in subsurface activities! If these business functions are not performed and if the company is not consistently profitable, the chances of it remaining in business are greatly reduced.

We will therefore require sophisticated management tools, and information systems that will allow us to manage our businesses better and enable us to assess risk in a more efficient manner. The improved quality of products will be vital in reducing exposure to risks by reducing costs and increasing profitability. These factors will provide incentive for all involved to conduct the necessary research that is not possible at the present time. Once this happens, the opportunity for innovative technologies will increase proportionally, and both consultants and contractors will share in the rewards and benefits.

Consultants are in most cases selected for design by technical merit with price being only one of the determining factors. This is a legal requirement under the Brooks Act. There are times when a consultant will be selected because of original ideas on how to proceed with a project, and occasionally this will involve innovative technology. The potential risk in trying a new technology and having it go wrong would have to be weighed against the possible reward in succeeding and being able to use this success to secure new business and widen reputations. The fact that in most public projects the work will be performed by the low bidder - one who may or may not have the necessary expertise - will increase the risk of problems being encountered in the construction of the project.

There are other issues of risk versus reward for the contractor that can provide more constraints than those of the owner or the consultant. In the Design-Bid-Build system the opportunities may be limited to value engineering. As noted above, profits in the industry are low and there is, in most cases, no substantial money available for research and development. Proven technology or equipment has to be brought in from a foreign country. However, the owner often has a reluctance to specify the product or technology because of concern that there will be lack of competition in the new technique. The rewards for the contractor to spend the time and money to develop or import this equipment or technology are therefore diminished.

Despite these constraints, there are times and projects on which new technology can and is being implemented. If the need is extreme for either owner or constructor, necessity will force the issue. This introduction can be achieved either in the pre-bid stage, by making the owner or the consultant aware of the technology, or sometimes through Value Engineering. When either of these methods is successful, the reward can be very satisfactory.

Different methods of project delivery systems, mainly in the bidding or procurement stage, are beginning to appear. Many of these increase dramatically the balance of reward over risk for the owner, engineer, and constructor.

4. ALTERNATIVE BIDDING PRACTICES

The ASCE specialty conference at Cornell University in June, 1990, dealt with the subject of "Design and Performance of Earth Retaining Structures". One section was, significantly, devoted to Contracting Practices. Nicholson's paper compared practices in Europe with those in the U.S., and discussed associated topics such as research and development, contract documents, liability, and the problems of vested interests. Of special interest, however, was the section dealing with the modifications and alternatives to standard bidding practices.

Nicholson (1990) noted it is common to find contract documents, including qualification clauses, which call for some limited review of a contractor's experience record by the owner or consultant. The owner's approval is (nominally) required before the specialty subcontractor may be employed by the prime. However, it is difficult to ensure that these clauses perform as intended: in a very competitive bidding atmosphere, the low bid prime contractor usually feels he has a "right" to use the subcontractor of his choice. This usually means he will choose on the basis of cost over experience, and rely on his own interpretations of the clauses to justify his selection. For example, certain "experienced" individuals can be hired temporarily, or materials or equipment suppliers can be engaged to furnish "technicians" to supervise certain more critical phases of the work. This state of affairs is another result of the low bid process, and clearly affords no incentive or encouragement to the innovative contractor.

A more attractive method has been the concept of prequalification, whereby only pre-approved contractors are permitted to bid to the prime. About one half of state highway departments currently are using or considering this method for certain types of work. Typically, though, prime contractors find themselves besieged by subcontract bids at the last minute from "new" companies claiming the suitable level of expertise. If the offer is sufficiently low, the contractor is tempted, and rarely does the owner intervene because he submerges his fine intentions in the self justification of "fair and open competition".

The good intentions of prequalification are further diminished by the fact that no national forum exists where standard guidelines are set, although, the Corps of Engineers and others are experimenting with a contractor rating program to prohibit contractors with "unsatisfactory" records from previous works from bidding other work. Each owner, however, typically has his own prequalification system, and tight bidding and award schedules rarely leave time for submitted references to be verified. A drawback of even rigidly applied prequalification is that it rules out the potential for contribution by the innovative specialist during the project's conceptual and design phases. This is everyone's loss, as the team is that much weaker.

The "Value Engineering Proposal" is a form of alternate proposal, long established in U.S. practice. Although more progressive in concept, it has found limited use in specialty geotechnical contracting. This is not solely because cost savings must be shared, rather it is because at bid time the prime contractor is typically unable or unwilling to assess the inherent risks. These risks include the fear that change may disrupt the work; the owner may not accept the scheme; and there may be insufficient time for approval. When presented with such concerns, often for little reward in return, most primes simply reject value engineered proposals.

In contrast, some of the most attractive avenues open to innovative contractors are part of the Design-Build concept, common in the bidding climates of Europe and Japan and promoted for many years by the FHWA. Design-Build allows the specialist geotechnical contractor to introduce cost-effective solutions that meet or exceed the owner's performance criteria. Design-Build contracting practices promote innovative design and accelerated construction, often with the use of equipment specially built for the purpose. The traditional role of the owner's representative - the design consultant - is often modified and may be expanded. The design consultant sets the performance criteria within practical limits and provides assurance that the owner's needs are satisfied. Review and critique of competitive proposals from specialty contractors and consultants employed by them ensures that the most economical solution is found. It can be applied in any project where the owner seeks an innovative cost saving solution to a particular geotechnical construction problem. There are four distinct options:

● **Post Bid Design.** The owner prepares a set of special design criteria (special provisions) that are included in the bid invitation and define the parameters for the alternate design. An owner-designed or "as-designed" system may also be included. After successfully pricing the project and obtaining a contract, the specialist then provides a design to the owner for review and approval.

Difficulty with this approach concerns the ability of the owner and contractor to agree on the design, after the award has been made. Disputes and delays may result, and often the contractor must modify his design, and usually compromise potential profitability. Also, to protect himself, the owner typically over-specifies the design parameters, and this will stifle innovation. However, this can be a very attractive option, especially for smaller, highly technical projects.

A recent example was the epoxy resin sealing and rebonding of an old concrete dam in North Carolina (Bruce and DePorcellinis, 1991). Although the scale of the work was relatively small, the situation was long standing, deteriorating and not amenable to resolution by conventional grouting methods. Given the wide range of contemporary drilling and grouting concepts - often proprietary - which were feasible, the owner was able to select his preferred technical solution while ensuring it was economically competitive.

o   Pre-Bid Design. Prequalified, selected specialists prepare designs for the owner's review prior to the bid. Approved designs become part of the bid package and the specialty subcontractor prepares a price to construct only his proprietary design. This method works best when the contractor is permitted to prepare plans of a conceptual nature only. Such plans exclude details which the contractor feels are unique to his design. So long as the supporting calculations address these details, the bid documents may include only enough information to make other contractors aware of the nature of the work. This is a positive opportunity for innovative contractors, who of course, must still remain cost effective.

An example of this approach has been described by Nicholson and Wolosick (1988) for the construction of a vertically post-tensioned caisson retaining wall in Atlanta, Georgia. In this project, conventional, inclined tie-backs could not be placed due to property right of way restraints. The contractor therefore had to be innovative in designing a solution to satisfy the performance criteria, while the owner again benefitted by being able to select the lowest responsive price.

o   Negotiated Work. The owner is committed to a team approach wherein the contractor becomes an important part of the team for all foundation and ground support aspects of the project. Risk sharing is integral: the contractor is responsible for the adequacy of the design and its construction, the owner is responsible for the accuracy of the information upon which the design is based. Costs are reduced, as the contractor includes less contingencies, and innovation is encouraged because the contractor is rewarded for economies of design and installation. And, of course, quality is enhanced due to the team partnering approach.

This principle was adopted for the recent sealing of seepage through the Left Abutment of Lake Jocassee Dam, North Carolina (Bruce, et al., 1992). An awkward seepage problem was described by the owner in contract documents and technical and financial proposals invited. Detailed discussions with responsive bidders were conducted, and the successful contractor was chosen on the basis of his perceived ability to respond in the field to provide a solution within the limits of the quantities of work originally foreseen.

o **Two Phase Bidding.** It many ways another type of negotiated bid, this has gained favor in recent years with many Federal and State agencies. Prequalified contractors are invited to submit separate very detailed technical and financial offers. The technical aims of the project, and special restrictions are clearly specified, but great scope is afforded to the inventive bidder. Each proposal is assessed independently by separate committees, and graded on a points system disclosed in advance. The value of the technical proposal typically exceeds that of the price proposal, and emphasizes technical competence, personnel and corporate experience, and safety. There may be successive "rounds" of bidding, with the responsive contractors being interviewed between times so that they can optimize their proposals to a "Best and Final" submittal.

During the negotiations, the successful contractor should have developed a full understanding of the requirements of the job, and so there should be no subsequent controversy over the specifications, scope of work, or the quality level intended. He may also not have the "low bid". Unsuccessful contractors will have incurred a great deal of bidding cost, but this prospect alone will deter all but the most serious contenders. This process also involves considerable effort on behalf of the owner, and so is really viable only on particularly large and/or complex projects such as the major repairs conducted at Fontenelle Dam, Wyoming, and Stewart Mountain Dam, Arizona (Bruce, et al., 1991).

Overall, therefore, there are significant opportunities for the owner and the consultant, as well as the contractor, in pursuing design-build options as opposed to the traditional low bid approach. These should be aggressively promoted throughout our industry. In summary, Design-Build provides benefits by
o   providing optimum solutions at the lowest possible cost;
o   encouraging innovation through contractor-sponsored research and development;

- fostering improvements in quality, performance and cost; and
- incorporating the most advanced and practical designs by prequalified contractors who are regularly and exclusively engaged in the business.

5. PARTNERING AND THE ROLES OF THE CONTRACTED PARTIES

The concept of partnering is not new - it has always been the fair basis of doing business between honest and responsible parties. In recent years, however, the formal process of partnering has attracted much favorable attention as an aid to avoiding conflict between contracted parties. Significantly, this initiative has now been fully endorsed by the Association of General Contractors, who have not always been so supportive of innovations in contracting or procurement procedures in the past. In September, 1991, they issued a booklet entitled "Partnering - A Concept for Success" which describes in detail the respective benefits to each party, potential limitations, and the basic mechanisms of the process.

It must be realized that partnering "is not a contract, but a recognition that every contract includes an implied covenant of good faith. While the contract establishes the legal relationships, the Partnering process attempts to establish working relationships among the parties (stakeholders) through a mutually-developed, formal strategy of commitment and communication. It attempts to create an environment where trust and teamwork prevent disputes, foster a cooperative bond to everyone's benefit, and facilitate the completion of a successful project" (AGC, 1991).

Recurrent themes in the respective benefits to each partner include reduced exposure to litigation, lower risk of cost overruns or project delays, enhanced quality and safety, and an "increased opportunity for innovation through open communication and element of trust, especially in the development of value engineering changes and constructability improvements". It is intended to be an opportunity in public sector contracting, where the open competitive-bid process keeps the parties at arm's length prior to award, to achieve some of the benefits of closer personal contact which are possible in negotiated or design-build contracts.

Regarding potential problems, it is fundamental that all parties fully commit to the concept. This may prove very difficult for those conditioned in an adversarial environment, or those who regard it as just another "fad", or those who thrive on daily confrontation and cannot entertain the idea of "win-win" thinking.

The AGC booklet (1991) provides a model to pursue on any given project, acknowledging that the details and personalities involved will necessitate appropriate "tailoring". Early steps in the process include educating one's organization in advance, making partnering intentions clear (from bid solicitation onwards), and committing and involving top management from award onwards.

Thereafter, a partnering workshop should be held before starting work on site, and in the authors' experience, this is an extremely valuable and interesting exercise. Preferably facilitated by neutral, trained specialists to aid focus, this workshop is an excellent opportunity to share concerns and establish common goals and objectives. It permits the team members to establish personal rapport, and should establish a clear communication framework. Partners can introduce specific proposals for innovation, both in the technical sense, and in others, such as an alternative issue escalation and resolution process. Ideally, the workshop will end with the creation of a partnering charter, signed at that time by each of the attendees.

This whole formalized process may appear to some to be somewhat superfluous, even leaving aside the question of the extra time and cost initially involved. Indeed, the objectives of partnering have often been achieved by reasonable people without recourse to such a structure. However, the authors view this formality as a necessary discipline to reinforce natural good cooperative instincts and personal relationships. Interim and end of contract evaluations are also an essential part of the charter, and help keep the concept in clear focus.

6. DISPUTES RESOLUTION

Typically in this process, an independent panel, paid jointly by the owner and contractor, meets on a regularly scheduled basis and is empowered to settle disputes among the owner, engineer and contractor as they occur. The keys to this are a knowledgeable board, a commitment by the owner, engineer, and contractor, and speedy resolution of problems as they arise. The board must have some power and authority, but usually this is limited to recommendations.

The greatest value of this process is forcing the parties to address the issues on a timely basis and with a neutral third party acting as a mediator. Once each party is forced to explain its own position and listen to the other party's position, a compromise solution can generally be achieved. In our present adversarial system, accusations and denials are issued in writing, without the benefit of a face to face meeting of the people who are decision-makers. If the party seeking additional money or a changed condition (typically the contractor) feels strongly enough about its position the dispute will escalate but only after many months or years have passed. In the meantime, positions will have hardened on both sides and the lawyers are called in for a legal resolution, either in court or formal arbitration.

Sometimes in addition to the Disputes Review Board, the contract will call for bid documents to be submitted separately but at the same time as the bid. These documents contain the contractor's estimate for the work and will show how the work was assumed to proceed. They are held in escrow pending a dispute that cannot be concluded by mediation before the Disputes Review Board. Either the Board or another arbitration authority will be authorized to open these documents and utilize their contents to determine intent and cost on the part of the contractor. This is a method of determining what was assumed to be the conditions at the time of entering into the contract and is a deterrent to the unscrupulous and a vindication of the honest.

7. FINAL REMARKS

For many years, the construction sector considered itself relatively isolated from the various forces impacting the American way of life, such was the strength of our domestic market and our power overseas. In recent years, however, intense competitive pressures have been brought to bear, as reported daily in the media, and experienced constantly by the participants. It has become a critical interlude in our corporate times, and one which will demand substantial change in attitudes and philosophies.

Already among contractors, consultants and owners there are encouraging signs that new approaches to contracting procedures, the concept of Partnering and alternative methods of disputes resolution are being promoted in a concerted effort to reduce confrontation and litigation.

Innovation and new technology must be addressed so that companies can become more productive, and responsive to clients. In this regard, there is an increasing awareness of the revived acknowledgement that quality is not just extra, quality is fundamental: the essence of TQM is that costs can be reduced and profitability increased.

On the other hand, there are several other factors which will probably remain outside our industry's collective ability to impact. Apart from imponderables such as Federal government fiscal policies, there will continue to be burdensome factors such as inflated insurance premiums and restrictive regulatory controls.

Nevertheless, it does seem that innovative and well managed geotechnical constructors can survive and prosper if they pursue the following basic goals:

- Innovate, by continually challenging technical and administrative paradigms.
- Recruit and train committed employees, empower them and remember that they are the company's most valuable asset.
- Understand and supply what the customer really wants.
- Promote alternative procurement procedures, especially design-build.
- Promote and embrace Partnering, team building, and non-litigious dispute resolution processes.
- Commit to total quality in every process of the operation, both internally and externally.
- Continue to innovate, in all phases of the operation!

In summary, risk and reward provide both the opportunities and constraints to the introduction of new technology. Under the present system, perceived and real risk on the part of the owner, engineer, and constructor have a dramatic effect on the acceptance of innovation. The team approach, rather than the present adversarial system is vital to successful change. It is essential to bring all the parties together during all stages of the project to work as a team to reduce risk and so gain the most economies. Equally, there must be a reward for those entrepreneurs who invest their resources into innovative technologies. Quality, not just minimum acceptable standards, must be maintained and rewarded.

ACKNOWLEDGEMENTS

The authors acknowledge with thanks the permission of the Associated General Contractors and Fails Management Institute to quote from their material. They also acknowledge the role of Col. Charles Cowan, currently of the Arizona Department of Transportation, in developing the concept of Partnering.

REFERENCES

1. Associated General Contractors of America (1991). "Partnering - A Concept for Success". AGC, Washington, D.C., 18 pp.
2. Baker, W.H., Cording, E.J. and MacPherson, H.H. (1983). "Compaction Grouting to Control Ground Movements During Tunnelling". Underground Space, I, pp. 205-212.
3. Bruce, D.A. (1988). "Aspects of Minipiling Practice in the United States". Ground Engineering, 21 (8), pp. 20-33, and 22 (1), pp. 35-39.
4. Bruce, D.A. and Nicholson, P.J. (1988). "Minipiles Mature in America". Civil Engineering, 58 (12), pp. 57-59.
5. Bruce, D.A. and DePorcellinis, P. (1991). "Sealing Cracks in Concrete Dams to Provide Structural Stability". Hydro Review, 10 (4), pp. 116-124.
6. Bruce, D.A., Fiedler, W.R. and Triplett, R.E. (1991). "Anchors in the Desert". Civil Engineering, 61 (12), pp. 40-43.
7. Bruce, D.A. (1992a). "Recent Progress in American Pinpile Technology". Proceedings of Specialty Conference on "Grouting, Soil Improvement and Geosynthetics". ASCE, New Orleans, LA. Feb. 25-28, pp. 765-777.
8. Bruce, D.A. (1992b). "Progress and Developments in Dam Rehabilitation by Grouting". Proceedings of Specialty Conference on "Grouting, Soil Improvement and Geosynthetics". ASCE, New Orleans, LA. Feb. 25-28, pp. 601-613.
9. Bruce, D.A., Luttrell, E.C. and Starnes, L.J. (1992). "Remedial Grouting Using 'Responsive Integration^{SM}' at Lake Jocassee Dam, SC". Ground Engineering, in print.
10. Business Week (1992). "Guilty: Too Many Lawyers and Too Much Litigation. Here's a Better Way". April 123, pp. 60-66.
11. Dobyns, L. and Crawford-Mason, C. (1991). "Quality or Else". Houghton Mifflin Co., Boston, 309 pp.
12. Engineering News Record (1991). "Top 400 Contractors". May 27.
13. Fails Management Institute (1991). Total Quality Management Short Course. Raleigh, N.C.
14. Nicholson, P.J. (1986). "In Situ Ground Reinforcement Techniques". Proceedings of Conference on Deep Foundations, Deep Foundation Institute/Chinese Institute of Ground Improvement and Sciences, Beijing, China, September, 9 pp.
15. Nicholson, A.J. and Wolosick, J.R. (1988). "Post Tensioned Caissons Resist Interstate Construction: A Case History". Proceedings 2nd International

Conference on Case Histories in Geotechnical Engineering, Univ. Missouri-Rolla, St. Louis, June 1-5, pp. 1425-1431.
16. Nicholson, A.J. (1990). "Contracting Practices for Earth Retaining Structures". Proceedings of Specialty Conference on "Design and Performance of Earth Retaining Structures", ASCE, Cornell University, Ithica, NY, June 18-21, pp. 139-154.
17. Peters, Tom (1984). In Search of Excellence. Harper and Row, Inc, New York, NY, 360 pp.
18. U.S. Bureau of the Census (1992). Construction Report, Series C-30, Value of New Construction Put in Place, Washington, D.C.
20. Warner, J.F. (1982). "Compaction Grouting - the First Thirty Years". Proceedings of Specialty Conference, "Grouting in Geotechnical Engineering", ASCE, New Orleans, LA, Feb. 10-12, pp. 694-707.
21. Warner, J.F. (1992). "Compaction Grout: Rheology vs Effectiveness". Proceedings of Specialty Conference on "Grouting, Soil Improvement and Geosynthetics", ASCE, New Orleans, LA, Feb. 25-28, pp. 229-239.

LIMEHOUSE LINK TUNNEL PROJECT
LONDON
A CASE HISTORY

Patrick McCreight[1], David Scott[2], George Tamaro[3]

ABSTRACT

The Limehouse Link road tunnel will connect the business centre of London with the new Docklands development in east London which includes Canary Wharf. This paper describes the need for the tunnel, the options available, the alignment, planning and design of the tunnel taking into account environmental and other constraints. Particular aspects of construction are covered including the utilization of slurry walls, top down method of construction, working through waterways, diversion of existing streets and utilities, monitoring techniques and use of river transport.

HISTORICAL BACKGROUND

London's Docklands stretched from London Bridge in the centre of the City over 10 miles along the banks of the River Thames. By the end of the nineteenth century the area had attracted a wide variety of trade. Wharves, warehouses and factories were served by a maze of narrow streets tightly packed with houses. The prosperity started to decline in the 1930s and the area received heavy bombing in World War II. From the late 1960s the docks closed one after another and the population declined as people moved away. It was obvious that the area needed special recognition.

1- Patrick McCreight is a Senior Technical Director and
2- David Scott is a Senior Executive Engineer of Sir Alexander GIBB & Partners Consulting Engineers of Reading, England a member of the LAW Companies Group of Atlanta, Georgia.
3- George Tamaro is a Partner, Mueser Rutledge Consulting Engineers, 708 Third Avenue, New York New York 10017.

In 1981 the UK Government set up the London Docklands Development Corporation (LDDC) to face the challenge of regenerating large areas of derelict land, encouraging private investment and raising the expectations of people living in the area.

Docklands is now recognised as a major opportunity area. For example the newspaper printing presses have moved out of the City of London (Fleet Street) into the Isle of Dogs. The major construction has been at Canary Wharf on the Isle of Dogs where Olympia and York have developed over 12 million square feet of office space which was occupied commencing in 1991.

The LDDC realised that an appropriate infrastructure was required to serve the area with good road and rail connections to the centre of London. The LDDC promoted the construction of the Docklands Light Railway system which now runs from Bank Station in the City of London through to the Isle of Dogs and Stratford to the north. The London subway system is being extended into Canary Wharf and on to Stratford. An extensive road system is being constructed of which the Limehouse Link forms an important part.

Fig. 1 shows the east end of London from the City to the Isle of Dogs where Charles I used to exercise his hunting dogs and indicates the location of the Limehouse Link Tunnel Project.

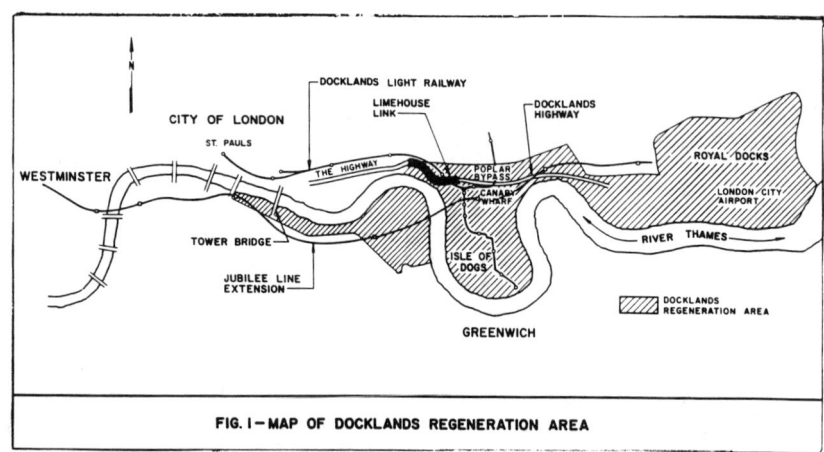

FIG. 1 – MAP OF DOCKLANDS REGENERATION AREA

THE NEED FOR THE LIMEHOUSE LINK AND ROUTES CONSIDERED

The road access serving the Isle of Dogs consists of narrow streets. A traffic model for the Docklands area was prepared by the LDDC and confirmed that these streets were totally inadequate for predicted traffic volumes. A Docklands Highway was conceived to improve communications for road traffic from the City of London to the Isle of Dogs and through to the east end of Docklands including the London City Airport. Traffic analysts concluded that the scheme should have sufficient capacity to allow new local traffic management schemes to be implemented and particular emphasis was given to the need to reduce traffic on local roads.

Fig. 2 shows the alignment of the Limehouse Link which commences at the east end of a street named The Highway running from the City and terminates at the Poplar Link forming part of the Dockland Highway. Ramps lead on to and off the Link at Westferry Road in close proximity to the Canary Wharf Development on the west end of the Isle of Dogs.

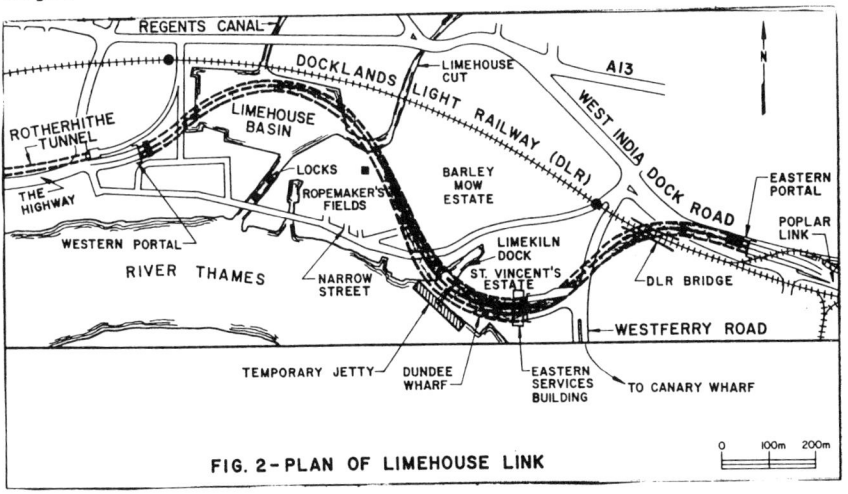

FIG. 2 - PLAN OF LIMEHOUSE LINK

The route is described in more detail below but it is of interest to discuss the alternatives which were considered at the planning stage.

The widening of existing roads was considered but rejected because of severe environmental problems including the demolition of many existing buildings. The river front of Limehouse along Narrow Street is of particular historic interest with many landmark buildings and is a unique survivor of 18th century Docklands.

A straight line surface route would have severe environmental disbenefits and would have had a major impact on the potential for redevelopment.

A tunnel constructed along the edge of the River Thames would have required significant demolition of buildings on the approach ramps to connect the tunnel to the existing roads.

ROUTE OF THE LIMEHOUSE LINK

From the road junction at the western end the route runs eastward on a descending ramp into tunnel and then curves along the northern edge of Limehouse Basin. The line of the tunnel is dictated by the presence of the approach ramp to the Rotherhithe Tunnel which was built at the turn of the century to provide access for horse drawn traffic under the Thames and is in constant use today. The depth of the tunnel in the Limehouse Basin is dictated by the navigable depth required for the connection between Regent's Canal and the River Thames. (Fig. 3)

The route runs south eastwards through an open area known as Ropemaker's Fields between existing housing (Barley Mow Estate) and other developments. The route then crosses under Narrow Street and runs under Limekiln Dock which is a tidal inlet from the River Thames. Again the level of the roof of the tunnel is governed by the depth of water required in the Dock.

The tunnel curves from east to north east through Dundee Wharf and St Vincent's Estate to pass under Westferry Road which connects to Canary Wharf. The tunnel widens from Ropemaker's Fields to accommodate ramps leading to and from Westferry Road. The tunnel continues to pass under the Docklands Light Railway (DLR) close to West India Dock Road. The crossing point was chosen in close consultation with the DLR Authority. The tunnel then crosses under West India Dock Road to rise into an open ramp to connect to the Poplar Link through to the east end of the Docklands area.

The total length of Limehouse Link is some 1.8km of which 1.5km is in tunnel.

The tunnel contains two carriageways with two lanes in each carriageway. The overall width is generally 24 metres but to accommodate the ramps leading to and from Westferry Road the maximum width is 48 metres. The maximum depth of the tunnel is nearly 20 metres adjacent to Westferry Road.

LIMEHOUSE LINK TUNNEL

Fig. 3 -   VIEW ALONG ALIGNMENT LOOKING EAST DLR VIADUCT AT LEFT, LIMEHOUSE BASIN RIGHT FOREGROUND, CANARY WHARF TOWER CENTER, REAR BACKGROUND

Four services buildings are provided. One at each portal contains mainly the ventilation exhaust fans with electrical transformers and control panels. A fourth building houses electrical equipment only. The building close to Westferry Road will also house the monitoring and control systems for the whole tunnel.

REGENERATION

When construction is complete, the re-established area above the tunnel will be available for redevelopment and plans are already in hand for this in Limehouse Basin where British Waterways Board, with a developer, have plans to build houses, apartments and offices along the northern edge of the basin. Walls are being built to suit the development which includes a marina for small boats. The area from Limekiln Dock to Westferry Road will have high development potential with a good river frontage.

Consequently most of the length of the tunnel has been designed to take the loads imposed by up to five story buildings sitting directly over the tunnel or alongside it. A layer of gravel minimum 3m thick will be placed over the roof of the tunnel to distribute loads from spread footings above the tunnel. Certain criteria have been drawn up to which the developers will be required to comply but there is no particular restriction on the position of future buildings.

PLANNING AND DESIGN OF THE TUNNEL

The design of the tunnel commenced in detail in January 1987 when Sir Alexander Gibb & Partners Ltd (GIBB) was appointed by the LDDC to confirm the route selection and to undertake the detailed design and, in due course, to administer the construction of the work on the job site.

At the commencement of the assignment it was proposed to invite bids for the construction of the project early in 1988 but this was put back to late 1988. The two year period for planning, investigations, design and preparation of the bid documents was extremely tight and meant that the LDDC and GIBB were working to a fast track program which required careful co-ordination with all parties involved.

It was apparent that the structural design of the tunnel would be largely dictated by the difficult soils strata in which the tunnel would lie. Two detailed geotechnical site investigation programs were carried out with 33 boreholes using sophisticated full length rotary coring and sampling techniques. It was revealed that the descending ground sequence was generally; made ground (clay, gravel, sand), alluvium (clays, sands and silts), gravel, London Clay (stiff but containing silt lenses), Woolwich and Reading Beds (waterbearing dense sands and silts with cemented layers) and Thanet Sands (very dense silty fine sand) overlying chalk at depth. (See Fig. 4) A great deal of effort was put into the interpretation of the results.

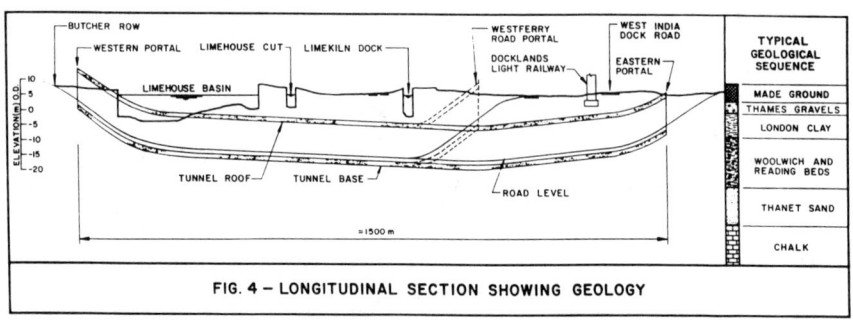

FIG. 4 – LONGITUDINAL SECTION SHOWING GEOLOGY

Studies were also made to collect historical and other data on defunct structures, wharves, jetties, quays, docks and the like which existed along the tunnel route. It became obvious that the construction technique to be used would have to be able to cope with a number of underground obstructions.

It was necessary to obtain extensive ground survey information. Rather than undertaking a full survey of the area which would have taken months, the availability of digitised Ordnance Survey maps made it possible to mount the information on GIBB's McDonnell Douglas GDS system. Plotting could be carried out to any scale to a high degree of accuracy. In order to supplement and update the OS data a limited amount of independent ground survey was also carried out.

GDS is compatible with MOSS software which was used in designing the link alignment. Merging of MOSS and GDS made it possible to see the route exactly and make alignment adjustments wherever necessary. This was useful in planning the utility diversions where drawings had to be circulated to the statutory authorities involved for their comments. When comments were received amendments could be made quickly and revised drawings sent out for approval.

Sophisticated finite element analyses using the latest techniques were carried out to model the complex soil/structure interaction for all stages of construction and the finished tunnel in both the short and long term.

The design of the tunnel is in accordance with the UK Department of Transport's requirements for trunk roads and tunnels in England and close consultation was kept with them. The design was independently checked under the Department's procedures for road and bridge works.

Structural condition surveys were made of all existing buildings along the line of the route. In order to minimise effects on existing buildings, it was decided that the corridor in which the tunnel would be constructed should be as narrow as possible. This placed a heavy emphasis in adopting construction techniques which could be used in a restricted corridor and with the minimum of working areas to either side.

From all these and other related studies, it became apparent that the "top down" method of construction was the appropriate solution. This comprised briefly:

> general site clearance with demolition of surface structures along the route
>
> construction of diaphragm walls (1.2m wide in 4.2m panels with an average depth of 20m) on each side of the tunnel
>
> installation of deep wells to reduce ground water pressure under the tunnel (as described later)

excavation between the external walls to roof soffit level with top propping of the walls where required

casting of the roof slab on the ground and dowelled to the diaphragm walls

excavation under the soffit of the roof slab to a depth of 9m for the base slab

concreting the base slab dowelled to the diaphragm walls

concreting a central wall to divide the tunnel into the two carriageways and concrete columns against the tunnel walls to provide full support to the roof

backfilling above the tunnel roof to the original ground level

Fig. 5 shows a typical cross section of the completed tunnel.

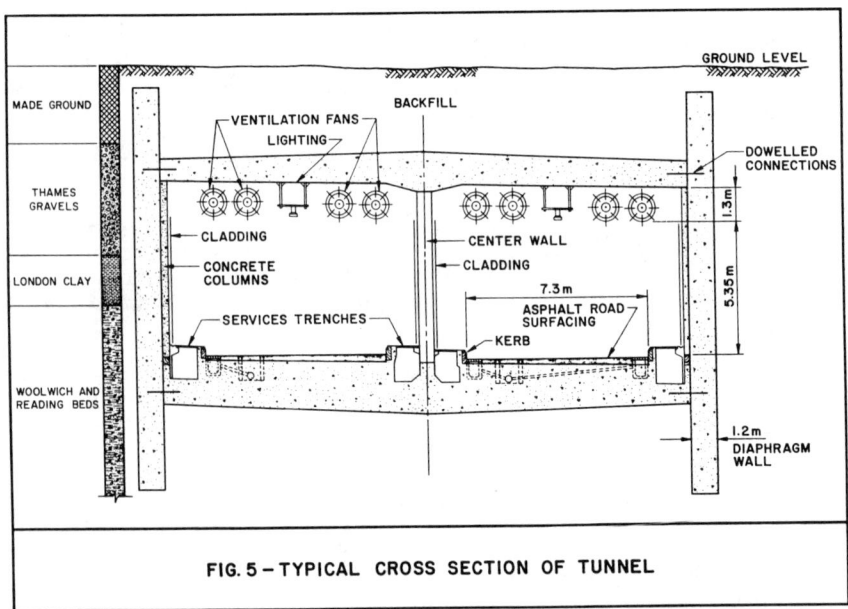

FIG. 5 - TYPICAL CROSS SECTION OF TUNNEL

Based on the method of construction stated above, detailed analyses were undertaken to predict the likely range of ground movements which would occur during construction and to decide which monitoring procedures would be necessary. The greatest movements were predicted within 10 to 15m of the tunnel where surface movements up to 20mm were expected. The monitoring systems used during construction are described later.

In parallel with the above, great efforts were made to locate the myriad of utilities which lay in the ground under roads and sidewalks on the tunnel route. Plans were made so that these utilities could be diverted temporarily during construction and then relaid back on the original alignments. This is also described in more detail later.

Plans were drawn up to maintain the flow of traffic in and around the area during construction particularly during peak hours and stages of construction were drawn up accordingly. Temporary bridges would be required where the tunnel crossed existing roads, for example Narrow Street and Westferry Road.

The area is within the London flood defence system which ensures that areas of London do not flood during high storm and tide levels on the River Thames. Planning had to ensure that the existing defence walls were not breached or, where breaching was necessary, adequate alternative defences would be constructed.

It was decided that a comprehensive condition survey should be undertaken of all buildings which lay along the edges of the route. From these surveys and the predicted movements which could occur during construction, recommendations were made and implemented for strengthening works which would be necessary to ensure that the buildings did not suffer distress during construction.

The LDDC advised that the land for the construction of the tunnel would be handed over in phases as existing buildings became vacant. The tunnel passes under the DLR to the east of Westferry Road and a new bridge had to be built by the DLR authority with two wide spans to accommodate the two carriageways of the tunnel. The timing of completion of the widened bridge had to be carefully researched and taken into account in the program for the tunnel.

It became very apparent that it would be necessary to plan all the activities which would have to be undertaken prior to construction commencing and during construction in order to enable the optimum period for construction to be determined. GIBB mounted 360 activities on a critical path network and researched widely through the industry to obtain rates of output for all the different operations which were necessary. This network was a valuable tool to ensure that the design and other related activities were co-ordinated effectively. The period for construction of the project was determined as 48 months.

Based on the envisaged construction techniques and the programme a number of studies were made to determine the effect on the environment both during construction and afterwards during operation. The studies included the impact of construction noise and vibration on existing properties along the line of the route, and the effect of traffic noise and the noise from the ventilation systems when the project was in operation. Air pollution was studied and, together with wind and climatic conditions, studies were made on the effect of traffic entering the tunnel and on the effect of the emissions from the ventilation chimneys. The effect of the impact on the environment during construction could be minimised by the erection of construction fencing and by the provision of sound insulation treatment to the worst affected properties along the line of the route.

It was necessary under the LDDC procedures to hold a public inquiry in order to justify the purchase of land which was needed for the construction of the project including job site working areas. At this inquiry the need for the project was justified and assurances made on the design and construction of the project. Evidence was given on the effect of the project during construction and in operation on the environment. This inquiry was held in October 1988 and the lead author of this paper gave the engineering evidence with reasons for the land required. The Inspector at the hearing upheld the LDDC's requirements with some minor modifications.

A great deal of effort was put into obtaining the exact requirements of the statutory authorities who had jurisdiction in the area. For example the British Waterways Board who own Limehouse Basin and the Regent's Canal had special requirements in regard to the use of the Basin and the lock gates. This information was carefully collected and included in the Bid Documents for the information of the bidders.

AWARD OF CONTRACT

It was decided that it would be appropriate to award a single contract for civil and building construction and to include in it the complete electrical and mechanical services and plant. Although it was fully appreciated that the contract would be won by a civil contractor and he would award sub-contracts for the buildings and services, it was decided that he should be fully responsible for all the planning of the interfaces between the different disciplines.

The bid documents were issued late in 1988 to pre-selected bidders and returned in May 1989. There was keen competition between six major companies.

One bid was returned by a joint venture of two major British civil engineering construction companies - Balfour Beatty Construction Ltd and Fairclough Civil Engineering Ltd (BBF). This joint venture offered an alternative form of construction for the 460m length through the Limehouse Basin with cost saving. The alternative comprised bottom up construction with a conventional cofferdam of heavy section steel sheet piling on both sides of the tunnel. The sheet piling would be propped with walings and excavation taken to the full depth. The base slab of the tunnel would be constructed first and then the side and centre walls built followed by the roof slab.

A detailed investigation of the alternative proposal was made and in particular the effect of driving the sheet piling on the environment and the effect of construction on the DLR brick arch viaduct which runs alongside the dock wall at the north end of the basin. It was decided that the scheme would be suitable for this particular length of tunnel which was remote from existing buildings. Accordingly the proposal was accepted.

Following negotiations, the contract for the construction of the project was awarded to the joint venture in September 1989 with a construction period of 45 months.

LONDON DOCKLANDS DEVELOPMENT CORPORATION

Prior to construction the LDDC established a working relationship with the local Borough in whose area the scheme would be constructed. Within this relationship which came to be known as the Accord, it was agreed that the LDDC would arrange for people living in buildings close to the scheme to be rehoused during construction and about 500 units were evacuated.

Other premises were eligible for noise insulation measures and approximately 500 units were fitted with secondary glazing to the windows overlooking construction.

Under the British system all works of major construction have to be examined by the local authority to ensure that working hours and noise levels are acceptable to the authority, in this case the Borough. Before the bids were returned provisional working hours and noise levels were established by the LDDC with the Borough. These hours and noise levels were not finally agreed until after construction commenced. Unfortunately the hours of working and noise levels became more restrictive than had been previously envisaged. This had an effect on the methods and times for construction which meant that a re-negotiation of certain aspects of the project were necessary a year after the contract had been awarded.

CONSTRUCTION

Construction commenced with clearance of the site in October 1989. Following the surveys of the existing buildings it was decided that it would be prudent to demolish a 14 storey block of flats which was close to the tunnel route in Ropemaker's Fields. The block had been built in the 1960s with a precast concrete design and comparatively shallow piled foundations. As a result the safety of the block could not be guaranteed even with the top down method of construction.

Construction commenced primarily in the areas of Limehouse Basin and Ropemaker's Fields with demolition of the two blocks of flats in St Vincent's Estate to the west of Westferry Road.

Particular aspects of the construction which are relevant to the topic of this paper are given in the following sections.

DIAPHRAGM WALLS

The construction of the diaphragm walls (slurry walls) was on the schedule critical path and was the key to the success of the early stages of the project. Aware of the importance of the diaphragm wall construction to the schedule as well as other aspects, the LDDC established a board of consultants to advise on geotechnical, construction and contractual aspects of the tunnel construction.

The diaphragm wall work was sub-contracted by BBF to three companies Stent/Bachy/Soletanche (SBS) acting in joint venture. The biggest perceived problem at the start of the project - the presence of ground obstructions at depth - did not materialise. BBF probed the whole length of the route to identify obstructions, including the possibility of unexploded bombs! Obstructions were cleared using breakers and backhoes and backfilled with cement bound material. In the area of Dundee Wharf, where an old timber drydock was present, mass excavation was required. Controlled blasting was used to demolish the walls of Limekiln Dock. Concrete guide walls were constructed to accurately locate construction of the diaphragm walls.

Steel reinforcement in the concrete in the diaphragm walls extended the full depth of the walls with a very heavy density to cater for ground pressures, the transfer of loads between the roof and floor slabs and for future development over the tunnel. Inserts for the dowelled connectors between the walls and roof and floor slabs had to be placed accurately. Stopends were provided at the end of each panel with dumb bell waterstops. In total 672 panels 4.2m wide were cast in depths up to 33m over a 20 month period.

Excavation in the bentonite stabilised trenches was carried out using crane mounted rope operated clamshell buckets. (Fig. 6) At one time four rigs were working on the site. Bentonite plants serviced two or more operations at one time and were moved along the job as the work proceeded. Tight tolerances were specified and achieved during construction.

Fig. 6 - CONSTRUCTION OF DIAPHRAGM WALLS IN TIGHT QUARTERS

The main drainage sump is located under the Eastern Services Building to the west side of Westferry Road. Here the diaphragm walling had to penetrate 9m into the extremely dense Thanet Sands with panels 33m deep. The Thanet Sands had been dewatered by an extensive depressurisation system which involved sinking wells into the underlying chalk strata. Flat, L and T shaped panels were required to a tolerance of 1 in 100. Handling the reinforcing cages in such great lengths required a high degree of planning and preparation and three heavy cranes were involved in lifting the cages.

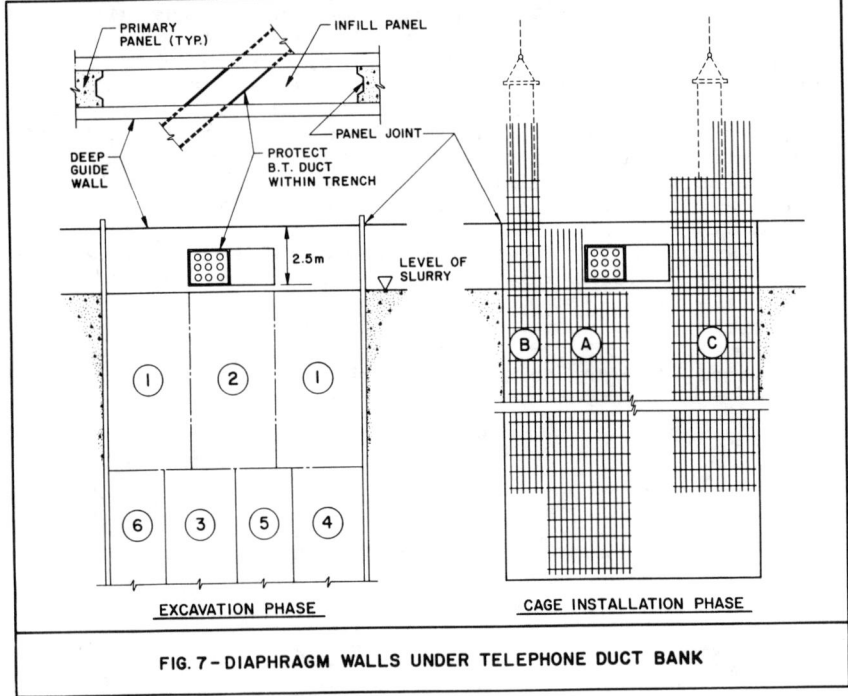

FIG. 7 - DIAPHRAGM WALLS UNDER TELEPHONE DUCT BANK

Construction Staging at B.T. Duct

1) Excavate Slot 1 to 8 meter depth

2) Excavate Slot 2 with backhoe

3) Excavate Slots 3 through 6 in sequence to panel bottom

4) Lower Cage A alongside Duct and slide beneath Duct

5) Lower Cage B into space between Cage A and joint

6) Lower Cage C (as in Step 4)

7) Lower Cage D (as in Step 5) (not shown)

8) Concrete panel with 2 tremie pipes

Large banks of telephone ducts crossed the site near Westferry Road at two locations and the diaphragm walls had to be carefully installed below them without disturbance. This was achieved by carefully spacing the panels with interconnecting half bites taken out from either side of the duct. The ducts were protected in boxes and by bars cast into the guide walls. The reinforcing cages had to be carefully installed working from both sides using an intricate pulley arrangement. A sketch of the technique used is shown in Fig. 7.

An interesting problem arose in the construction of the diaphragm walling underneath the DLR bridge to the east of Westferry Road. The headroom from the top of the finished wall to the underside of the bridge was only 6m. This was governed by the fact that a deep excavation was not permitted under the bridge to avoid the risk of settlement. SBS adapted two clamshell buckets to a closed height of 4.5m and a crawler crane was modified to operate with a stub jib. The walls had a depth of 18m and the cages were split into four sections for installation. The positioning and placing was handled by a gantry running on rails set along the guide walls. Cage distortion during building and coupling was potentially a problem and the cages were constructed on carefully prepared level beds. Precast stopends were used with tubes a manchette incorporated within the stopends for subsequent grouting at the joints. Planning of this particular operation had to be undertaken very carefully and contributed to the success of the technique.

A photograph of the rig used at the DLR crossing is shown in Fig. 8.

Low Head Room Rig   Cage Installation

Fig. 8 - DIAPHRAGM WALL CONSTRUCTION AT DLR CROSSING

The diaphragm walls have now been exposed in the tunnel and the quality of the work and the accuracy of casting the panels has been very good. Without doubt the success of the project has lain in the concept and execution of the top down method between the diaphragm walls.

TOP DOWN METHOD OF CONSTRUCTION

As stated earlier in this paper the top down method of construction was specified for the tunnel as this was considered to be a safe operation. It also permitted construction to be carried out under the roof slab with less likelihood of exceeding the onerous noise limits.

Before construction started an extensive system of wells was constructed to lower the water table to a safe level below the tunnel formation level to prevent risk of heave of the formation with the possibility of danger to the tunnel structure and surrounding property. Depressurisation was achieved by a series of pumped wells installed outside the diaphragm walls to the underlying permeable layers (such as Thanet Sands) as construction proceeded. Drainage of the overlying beds inside the diaphragm walls was assisted by sandwick or PVC vertical drains which toed into the Thanet Sands. At a few locations wells were necessary within the cut to alleviate particular problems. The base slabs and the slab in the deep dewatering sump have all been cast on a satisfactory dry foundation.

BBF used steel tubular props up to 1.3m diameter with steel walings mounted on concrete corbels pinned to the diaphragm walls above the roof level. The roof was cast with access holes approximately 15m long by 4m wide spaced every 50m. Through this hole small bulldozers and backhoes were lowered into the tunnel. A pilot tunnel was excavated from access hole to access hole using an extract ventilation system. The pilot tunnel was then enlarged and excavation continued to the full depth of the tunnel. (Fig. 9)

Extensive monitoring of the diaphragm walls was installed to observe inwards movements. Safe limits were calculated by analysis of the pressures arising from the surrounding ground and nearby buildings. Emergency measures were available to introduce steel props across the tunnel above the floor slab level if inwards movements exceeded the calculated movements. Once the excavation had been completed to full depth, the reinforced base slab was cast.

Fig. 9 - EXCAVATION "UNDER THE ROOF"

LIMEHOUSE BASIN

The tunnel construction passes through the northern edge of Limehouse Basin. The basin was excavated in the 18th/19th centuries and was an important coaling basin through which coal passed from the South Wales coalfields to the City of London. The Regent's Canal runs from the northern side of the Basin up through London and into the canal systems to the north of London. Limehouse Cut on the eastern side of the Basin is connected to the River Thames on the River Lea and acts as a discharge for storm water which passes down the Canal. New lock gates funded by the LDDC were built before construction of the tunnel started so that the contractor could have access for river barges into the Basin for bringing in construction materials and removing spoil.

At the commencement of construction through the Basin it was necessary to remove by dredging approximately 125,000 cubic metres of debris and silt some of which was heavily contaminated with metals such as arsenic, cadmium, lead and mercury which had to be taken to special disposal areas. Once the silt had been removed a strip of land along the line of the tunnel was brought up to flood level by depositing marine fill dredged from the North Sea. The fill was transported up the River Thames in barges and pumped hydraulically through a pipeline to the northern edge of the Basin where it was levelled by bulldozers. This formed a level platform on which the heavy steel sheet piles and tubular piles for the cofferdam could be set up and driven. (Fig. 10 and 11)

Fig. 10 - SHEETPILING IN LIMEHOUSE BASIN

This particular part of the project suffered some delays in that the sheet piling was more difficult to drive than expected and the noise from the pile driving hammers travelled across the water to a new development which had been built in the previous two years adjacent to the lock. A delay occurred until the residents in the new flats received a double glazing package or compensation in lieu.

Construction of the tunnel commenced adjacent to the Regent's Canal which was closed. Arrangements were made to allow storm water flows to pass from the canal into the Basin. Once this section was complete, the Canal was diverted back over the tunnel and construction commenced across Limehouse Cut. During this period it was necessary to provide a pipe across the cofferdam to allow flood flows which could enter the Basin from the Canal to pass into Limehouse Cut.

DIVERSION OF ROADS

The tunnel alignment crossed two major roads which had to be bridged to allow construction to proceed without hindering traffic. At both crossings Bailey Bridges were erected which have been a standard military bridging system since World War II. The bridge at Narrow Street was 53m long and at Westferry Road 35m long. The bridges were erected adjacent to the existing roads and the roads were then diverted on to them. A small bridge was also erected just east of Westferry Road. The erection of the bridges meant that the Contractor had a through access for materials and plant along the tunnel route. Utilities were suspended from the bridges as described below.

Fig. 11 - LIMEHOUSE BASIN CUT AND COVER

DIVERSION OF UTILITIES

During the construction of the tunnel particular attention had to be given to the diversion of statutory authority utilities at the road crossings. This area of east London has been the subject of urban development, probably several times since the start of the industrial revolution and utilities have been laid and relaid on numerous occasions. In some cases records of the locations of the utilities were sparse.

Diversions were required for the following utilities:

Electricity supplies - Extra High, High and Low Voltage
Gas supply, medium and low pressure mains
Water supply, mains and feeders
Sewerage, local and trunk routes, foul and surface water
Telecommunications, British Telecom, Mercury Communication (many utilities laid in old hydraulic pipes) and cable TV.

The brief was to divert the utilities in advance of the tunnel work programme and maintain the utilities during construction of the tunnel. All the above utilities were present at most locations but certain interesting features at the major locations are covered in the paragraphs below.

At the west end of the tunnel at the junction of The Highway and Butcher Row, the utilities were diverted away from the tunnel works while maintaining traffic flow through the junction. A ventilation access shaft from the Rotherhithe Tunnel which runs under the River Thames was found to be on the line of the new works. The shaft was redesigned and extended to the requirements of the Borough and the Department of Transport.

Under the road which runs down the western side of Limehouse Basin, there is a major telecommunications link from the City of London to the East London Satellite Link. This comprised both fibre optic cables and copper pairs in a deep duct route. During the design stage it was intended that this should be supported together with other utilities on a bund constructed in Limehouse Basin. Due to the long time required for jointing the copper pair cables and due to the fact that the construction of the bund had critical programme implications, it was decided to divert the cables permanently to a new route.

In Ropemaker's Fields which had originally been used for recreational purposes it was necessary to provide two footway bridges over the tunnel works. Existing utilities were diverted over the tunnel on hangers under the bridges. Adjacent to the area was an old, non-operating power station. However part of the yard was still used as a transformer station. As a result more than sixty old cables were found along the route of the tunnel. It was necessary to prove that the cables were dead by "spiking" them, generally using a cartridge gun device for personnel safety.

Further down Ropemaker's Fields a footway had been used as a route for the Extra High Voltage (EHV) supplies to the transformer station. The electricity authority was planning a change in the supply routing programme. By careful planning, liaison and good fortune the diversions were integrated with the authority's plans.

At the crossing of the tunnel with Narrow Street the utilities were diverted and laid in a steel box at grade on the line of the proposed bridge. The bridge was constructed over the top of the box and the box was then suspended by hangers from the bridge. This avoided the possibility of serious delays if the utilities had been hung from the bridge after erection.

Further along in the Dundee Wharf area a temporary substation was built in order to remove an existing sub-station. EHV supplies were diverted over the tunnel on a bridge.

At Westferry Road, British Telecom were installing two 48 way duct routes to feed the Canary Wharf Development and, in co-operation with them, the duct routes were boxed ready for suspension from the bridge at a later date. The diaphragm walls were built around this duct as described above.

At the West India Dock road which is a major route into the Isle of Dogs extensive utilities were found and identified. The utilities were diverted over a section of the tunnel roof which had been built earlier. British Telecom ducts were suspended on a support system to reduce the delay due to jointing and rejointing which would have resulted if they were diverted. Pipe jacking was used to divert a sewer as a trench in open cut would have been difficult to construct with a high water table and difficult ground conditions.

At the eastern end the link fits into road works designed as part of the through route and compatibility was achieved with the utilities laid under the adjacent contract.

The extent of work needed to identify and plan the diversion of utilities cannot be underestimated particularly when the utility companies have to be mobilised to come to the site. Delays in this type of work can have serious repercussions on the programme achievement of the more expensive civil contractor. An appropriate solution had to be found and implemented at each road crossing.

INSTRUMENTATION MONITORING SYSTEM

An extensive instrumentation system was established to monitor the response of 40 existing structures along the route and the tunnel structure itself during construction. Depressurisation of the deep aquifers by means of wells in the Thanet Sands and overlying Woolwich and Reading Beds was monitored by a large number of piezometers.

Existing structures varied from a 14 storey residential block, constructed of prefabricated reinforced concrete panels and founded on a piled raft, to the 150 year old brickwork viaduct carrying the DLR near Limehouse Basin. The DLR bridge running over the line of the tunnel was also carefully monitored. The techniques used to monitor vertical and horizontal movements and rotation are described below.

Stainless steel sockets were installed on each structure to allow levelling pins to be fitted when required for precise levelling to monitor total and differential vertical movements. Survey control for the precise levelling was provided by four deep datum points (sleeved fiberglass rods) drilled into the underlying chalk at depth. Shallow reference points could not be assumed to be stable because of the settlement due to groundwater depressurisation.

Measurement of horizontal strain within the structures was carried out using tape extensometers (steel tape with tensioning device) and eyebolts fitted in the sockets. Total horizontal movements were determined by the co-ordination of a small number of sockets, principally at the corners of buildings. Inclinometer tubes were installed in the ground where structures were founded on piles to monitor horizontal movement with depth. Rotational movements were monitored by electrolevel devices fixed to the faces of structures.

The performance of the tunnel structure was monitored in terms of wall deflections and loading in the temporary props placed above the tunnel roof. Inclinometer tubes were cast into the walls at various locations along the length of the tunnel and in some cases extended below the toe of the wall into the Thanet Sands to provide a fixed reference point. The temporary top props were required during excavation from ground level to roof slab formation level at certain locations to limit wall movements adjacent to existing structures. Prop loads were monitored using vibrating wire strain gauges to ensure that the capacity of the tubular steel props would not be exceeded. Elsewhere top of wall movements were measured by survey techniques. During excavation to base slab formation, wall movements (convergence) were monitored by distance measurements with tape extensometers.

At three specific locations along the tunnel a full instrumentation section was installed to provide data for an enhanced understanding of the tunnel structure performance and to permit back analysis. A typical arrangement of instruments including inclinometers and electrolevels in the walls, push-in pressure cells close to the walls and piezometers is shown in Fig. 12.

Construction of the tunnel required the depressurisation of two main aquifers, namely the sand lenses in the Reading and Woolwich Beds and the Thanet Sands. A number of standpipes and pneumatic piezometers were installed in these strata to monitor drawdown of the piezometric elevations as excavation progressed. The effect of construction activities on the level of the groundwater in the near surface Thames Gravels was monitored using piezometers.

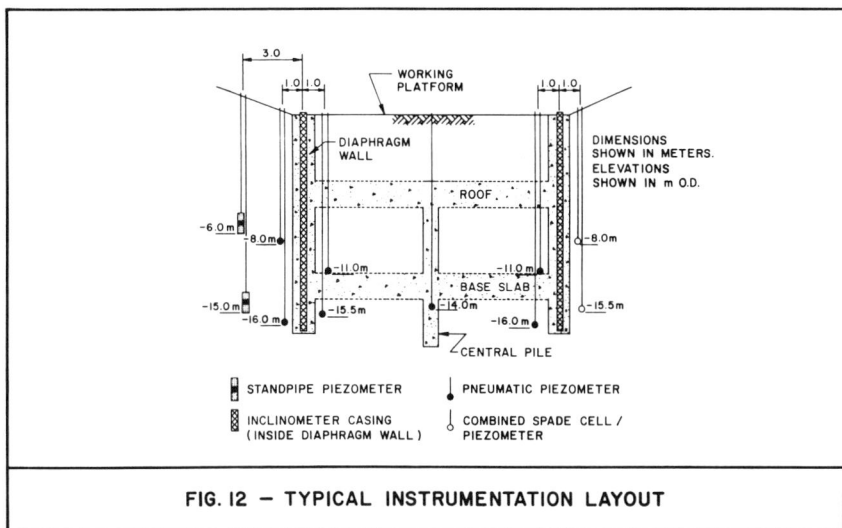

FIG. 12 — TYPICAL INSTRUMENTATION LAYOUT

Monitoring of the instrumentation was carried out daily by the contractor and GIBB. A system was established whereby monitoring data was compiled daily on computer disk in the form of LOTUS 123 spread-sheets. The data was examined, assessed daily and plotted in graph form to show movement trends and to identify the need for early action using the three level classification system defined below:

green level   stable readings within design predictions

amber level   readings exceeding the design predictions indicating possible problems

red level   readings at or above the level where structural distress and/or instability could occur; action required to halt or reverse trend of behaviour.

Typical results have indicated that structures adjacent to the tunnel experienced total settlements of up to 20mm which were largely due to the groundwater depressurisation. Differential settlements were generally of only 4 or 5 mm.

Horizontal movements of the buildings were also very small (less than 3mm). Maximum deflection at the top of the tunnel walls was 15mm and prop loads rarely exceeded 80% of the design load.

It is intended that monitoring of the structures should continue at regular but less frequent intervals for a year or so after the tunnel has been completed in order to ensure that there is a full record of behaviour in the area before, during and after construction.

TRANSPORT OF MATERIALS BY RIVER

Arising from the agreement reached between LDDC and the local Borough in the Accord, it was a requirement that the contractor should make maximum use of river transport to bring materials to the Site and to take surplus spoil off site. The clear intention of this was to reduce the movement of vehicles on the local roads which are heavily congested at peak periods. The requirement to maximise barge movements naturally had a cost consequence which was reflected in the bid but the LDDC considered that this was an acceptable price to pay for the clear environmental advantage.

The contractor decided to erect a wharf 200m long on the edge of the River Thames to accommodate barges bringing in cement, pulverised fly ash, aggregate and sand and other materials. The aggregates were dredged in the North Sea and processed at Dagenham further down the River Thames and brought to the site in 500t barges. The aggregates are stored in large hoppers and fed to twin concrete batching plants with a capacity of 190 cubic metres per hour erected on the back of the wharf (Fig. 13). The concrete is pumped through a 125mm diameter pipeline to both ends of the job. Truck mixers are used to augment the concrete pipeline for large pours.

A fleet consisting of more than 60 barges, lighters, tugs and other craft was utilised to move 800,000 cubic metres of spoil from the excavation of the tunnel. The material is taken down the river a distance of 22 nautical miles to a jetty which has been built to receive the materials. The operation went smoothly and avoided delay to the construction programme. The wharf is being operated in reverse to bring onto the site marine fill for placing over the roof of the tunnel. 5,000 tonne barges are delivering 1,000 tonne of fill per hour by conveyors.

Some of the spoil was removed from the Limehouse Basin but the Basin has a lock entrance and the size and frequency of disposal is governed by the size of the lock.

Fig. 13 - EXCAVATION AT DUNDEE WHARF, CONCRETE PLANT ON WHARF AT TOP

Across the River Thames at Greenwich a large yard was set up to cut and bend the reinforcement required for the job. This was loaded onto barges and brought across the river to the temporary wharf. General goods are handled through a commercial wharf at Purfleet further down the River Thames which is conveniently situated near the orbital motorway (M25).

Without doubt the convenience of river transport has been an asset for the project and the contractor is to be commended for providing a first class facility in this respect.

CONTRACT COMPLETION

The opening of the tunnel is currently scheduled for May 1993 which is a period of 6.3 years from the commencement of detailed design. Following opening there will be a period for final restoration and rediversion of the utilities back to their final locations. The temporary wharf will be removed and the river wall reinstated.

The success of the project has relied on a number of activities starting with thorough planning, attention to the environment, wide consultation including the local community and the construction skills of a competent contractor and his sub-contractors. Without doubt completion will mark a milestone in the development of regeneration in the Isle of Dogs and further east and should stimulate further interest by developers.

Acknowledgements

Thanks are due to the London Docklands Corporation for their permission to publish this paper and to our colleagues and staff for their contributions and help during the preparation of this paper.

The Reconstruction of the
Morton Street Evacuation
and Ventilation Shaft

Daniel M. Hahn, P.E.

Abstract

This paper describes the geotechnical investigation, design and construction of a shaft to an operating tunnel on the PATH system in New York. This work was accomplished without interrupting train operations and without incident.

Introduction

The PATH Rapid Transit System connecting New York & New Jersey has recently undergone a major rehabilitation and upgrading. This program, which cost almost one billion dollars, included the rehabilitation and/or reconstruction of a number of ventilation shafts, stations and the upgrading of several electrical substations. Mueser Rutledge Consulting Engineers was involved in several of the ventilation shafts and reconstruction of subway entrances. The one which will be examined in this paper is the Morton Street Ventilation and Emergency Exit Shaft. Construction of this facility involved a deep shaft through soil and a connection into an existing railroad tunnel built almost 100 years ago. This work was done without interruption to train operations.

Background

The original construction of the tunnels, then called the Hudson Manhattan Tubes is a fascinating story in itself. The Morton Street shaft played a small but important role in this story. First of all this shaft, which is immediately adjacent to the Hudson River on the New York City side played no part in the construction of the under river tunnels. Rather it was used as the starting point for the New York land tunnels which would eventually terminate at 33rd Street and Sixth Avenue. This site was chosen because of its close proximity to the Hudson River thereby assuring quick, easy and inexpensive delivery of materials to the construction site. This also was one of the sites of the power plants used for construction. The plant generated 40 kw of electricity and 1200 boiler horsepower. Its compressor plant could generate 10,000 CFM.

PATH's objective for this project was to improve safety for its approximately 190,000 passengers who daily ride this rapid transit interstate rail system. The proposed rehabilitation of the Morton street Emergency Facility Shaft would provide greatly improved ingress and egress and ventilation for the tunnels under and near the Hudson River in emergency situations.

The existing Morton Street emergency exit, along with its two outdated, out-of-service and undersized tunnel ventilation fans, was constructed in the early 1900's. The kiosk, (See Fig. 1) which housed the exist shaft was located at street level about 75 feet from the Hudson River bulkhead. The exit was comprised of a narrow (2'-6" wide), steep, and winding steel stair (See Fig. 2) which extended from the surface to the track level of the under river tunnels 65 feet below. (See Fig. 3).

The inadequacy of the existing ventilation and egress system was dramatically demonstrated by the occurrence of a fire on March 16, 1982 in the tunnel in close proximity to the Morton Street Shaft. There was a heavy smoke condition which created difficulties for the Fire Department entering and Patrons existing the tunnels. Approximately 400 passengers had to be evacuated to the surface via the existing Morton Street egress shaft. That narrow stairway marginally served its purpose. Although several persons were hospitalized there were fortunately, no fatalities. Nevertheless, the fire reinforced the need for improved ventilation as well as better ingress and egress to and from the tunnels for both passengers and emergency personnel.

The new Morton Street Emergency Ventilation Facility is situated between the Hudson River bulkhead and West Street, roughly opposite Morton Street. The project entails the construction of two ventilation buildings with a center plaza. The two buildings would each house a ventilation fan and emergency ingress/egress stairs, one for each of the tunnels.

In determining the precise locations of the shafts for the project, the following criteria were considered:

Figure 1. Photo of Kiosk which Housed Shaft

Figure 2. Photo of Exit Shaft Interior

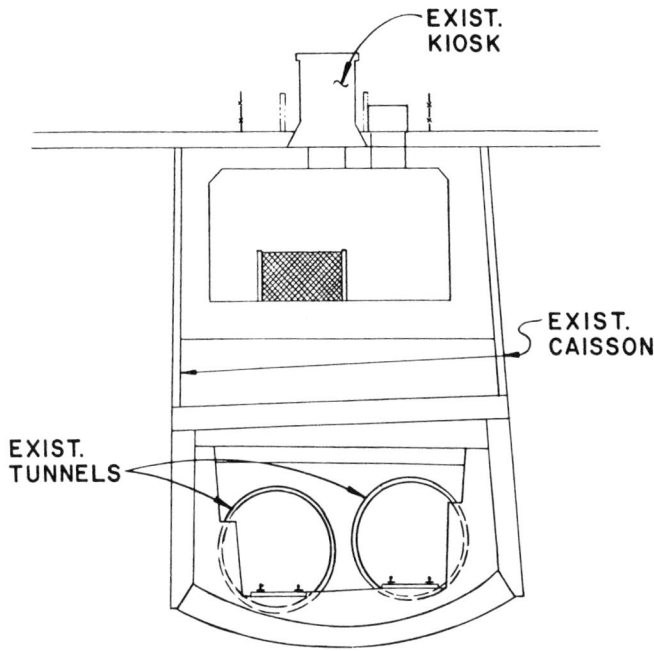

Figure 3. Profile View of Tunnels and Shaft

- The potential for utilizing a site which was either PATH owned or had surface or subsurface easements, or where an exit or ventilation structure presently existed.

- The facility was to be located at a point closest to the ends of the PATH tunnels under the Hudson River (approximately 1 mile in length) that would facilitate tunnel ventilation and passenger evacuation in the event of an emergency.

- Avoid potential negative impacts to the structural integrity of the tunnels by breaking into them at existing caissons rather than other locations.

- Making the break through into the tunnels as simple and as quick as possible which would be the least disruptive to operations and therefore the public.

It should also be pointed out that the shafts close proximity to the Hudson River posed a potential threat to the tunnels and PATH patrons, if through an error or an accident water from the river entered into the tunnels. Any overtopping of the new shaft walls next to the Hudson River would permit major inflow of water to the PATH tunnels. Since the Hudson River and the lower bay of the Hudson provided literally an infinite source of water, flooding of the tunnels at Morton Street could proceed through uptown tunnels to 33rd Street where water could then enter the New York City subway system. Likewise any water entering the shaft could also proceed toward New Jersey, flooding those tunnels and proceed back to New York through the downtown tunnels thereby flooding the basement of the World Trade Center. The construction at Morton Street had to proceed with extreme caution and the construction methods had to be fail safe. Needless to say, the new walls surrounding the tunnels were brought up to an elevation sufficient to protect the railroad from any infiltration of water which might arise due to exceptionally high tides, storms, hurricanes and any other natural phenomena that could be considered.

Subsurface Investigation

The subsurface investigation for this program involved approximately 10 borings which extended down to the top of rock. The first 20 foot below existing grade indicated that the material was fill material and contained cinders, sand, brick fragments, wood, granite paving block, asphalt, concrete, etc. The next 20 feet or so, was comprised of primarily river silt. Obviously the silt was formed prior to the filling in of the river in this area. The next 50 feet or so the geology was a series of sands, grayish brown, and brown with traces of gravel and/or silt. Below this strata approximately 90 feet from the surface, rock was encountered. The rock was a very competent mica schist and this would be the strata upon which the slurry walls would be founded.

The second part of the subsurface investigation is somewhat unusual. The existing caisson built in the late 19th century, was somewhat of a mystery, as there was very little recorded information about this structure. We knew that it was built of timber and brick but the exact dimensions and thickness of material were unknown. Thus, an investigation by boring through the caisson walls from the inside of the caisson was undertaken. Descriptions of materials encountered were somewhat unusual for a geotechnical investigation. For example, one boring produced the notation that seven foot from the interior face of the wall was a "waterproof seal", another boring indicated that "wood (vertical grain)" was discovered. A third boring indicated the following: "concrete, trace of wood, trace of glass." As a result of these investigations we had a good indication of what we were confronted with concerning both the geology of the site and the construction and geometry of the existing caisson.

Design

The existing access to the tunnels at Morton Street as previously indicated was via a caisson built during the time the tunnel was constructed. This is shown in Figure 3. The project involved substantially expanding the ingress/egress to and from the tunnel for emergency purposes, providing additional ventilation capacity through much enlarged shafts and the construction of a below grade space for an electrical substation. (See Figures 4 and 5). Simply put, the design approach was to construct two shafts A and B down to track level using slurry wall technology. These slurry walls would be constructed down to rock which would provide a groundwater cutoff as well as a foundation for the ventilation building. The area of slurry wall for this project was 50,000 s.f. of which 28,000 s.f. was for the ventilation shaft construction. In addition, the substation would be constructed also using slurry walls down to rock, however,

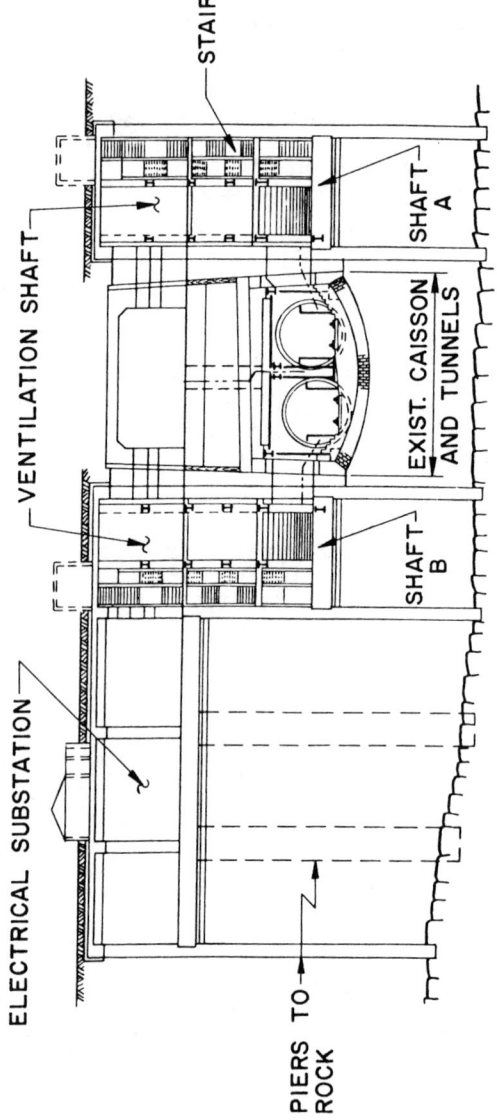

Figure 4. Profile View of Site

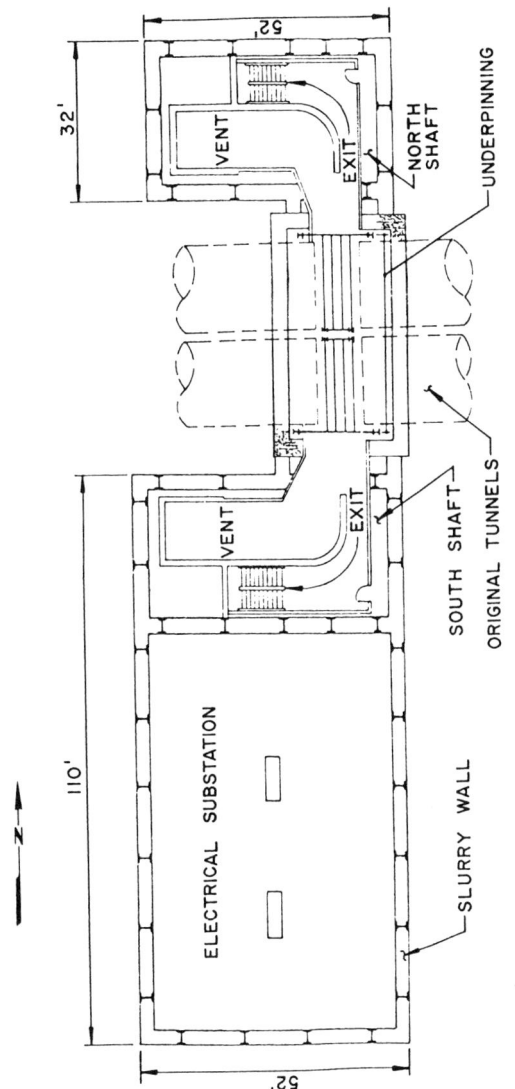

Figure 5. Plan View of Site

the base slab of the transformer room would be substantially higher than those for Shafts A and B. While not to understate the design approach or slurry wall technology, the key to a satisfactory completion of the project was the staged construction approach that was prescribed in the contract drawings. This staged approach was absolutely necessary in order to maintain a safe environment for the transportation of 190,000 people per day, through the tunnels. Any design or construction errors, or adverse environmental events could have had a disastrous effect on transit operations. Accordingly, the first stage of construction was to excavate trenches for the guide walls for the slurry wall panels. This is shown in Figure 6. In addition, the preparation of the caisson for underpinning was also accomplished in this stage. Grouting to consolidate the soil between the bottom of the existing fan room, the top of the existing caisson and the areas above the existing arch, was accomplished. Rock bolts were installed in the arch immediately above the tunnels. The next operation involved excavating for the slurry wall panels (See Figure 6) and installing these panels. At the same time that the slurry walls were being installed, underpinning of the arch above the tunnels could also commence. The next step (See Figure 7) involved the excavation for the substation and the bracing of the slurry walls. When the excavation was completed to subgrade, the mud slab and stone backfill was installed as well as the tiedowns and base slab construction. Tiedowns were necessary as an economical means of preventing the substation, (in essence, a basement,) from floating as the groundwater elevation was the same as the Hudson River. The installation of the tiedowns proceeded without incident except for two holes which had some minor water leaks. These were grouted. When the construction of the substation base slab and walls, along with the excavation in Shaft B down to the base slab in the substation were complete, work in Shaft A could commence. The procedure for construction in Shaft A to first excavate and remove the guide walls and excavate to 2 feet below the first bracing level, remove the water within the slurry wall limits and install cross lot bracing. This procedure was repeated five times until the excavation down to approximately 50 feet below the street surface. Space within the slurry walls was continuously dewatered and at this time a mud slab, stone backfill and base slab with tiedowns were constructed at the bottom of Shaft A. You will note from Figure 7 that there is a distinct clearance between the outside face of the existing caisson and the outside face of the new shafts A and B. At this time horizontal core borings were taken through the wall of the shafts to determine if grouting was necessary to make the

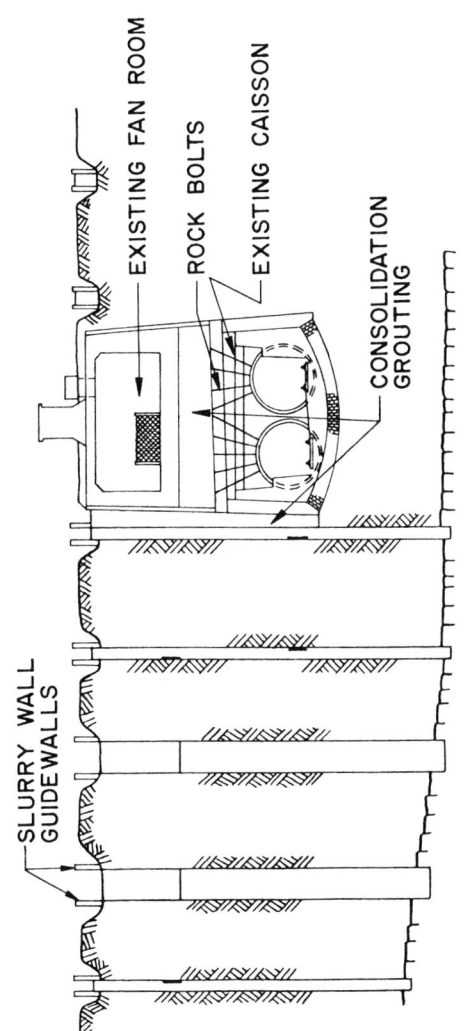

Figure 6. Profile View of Slurry Wall Construction

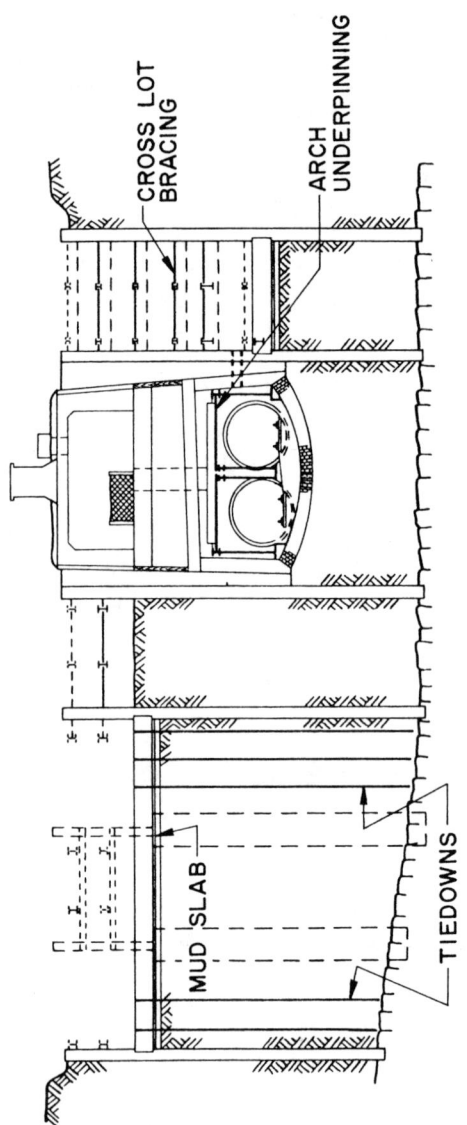

Figure 7. Profile View of Excavation and Bracing

shaft wall - caisson interface watertight, and also to verify the shaft location and tunnel alignment. After the grouting between the caisson and the shaft was completed, making a watertight seal, a hole was cut through the wall of the shaft. At the same time, a hole was cut through the wall of the caisson and the caisson wall was underpinned. Special procedures were used throughout the cutting of the caisson and underpinning. The contract specified twenty two steps of distinct work in connection with major relocations of traction power cables, compressed air piping and signal and communications systems. While the interior stairways, ventilation shafts and deflectors were constructed in Shaft A, the excavation and internal bracing of Shaft B was commenced. (See Figure 7). The excavation of Shaft B and the connection to the caisson proceeded as described for Shaft A. Now that shafts A and B were complete, and they were connected to the caisson, the only work left was finish work in both shafts and the substation. In addition the existing fan room had to be modified and the existing access to the street removed.

An extensive monitoring program was established so as to determine, for both the existing and new structures, displacements and stresses. Specifically stresses and movements in the existing structures and tunnels, vertical and horizontal soil movements around shafts and bulkheads, optical observations of wall and brace movements as well as piezometric water levels were observed and analyzed. The results of these observations revealed minor movements. Surface ground movements measured a maximum of ¼ inch. Lateral movements measured by slope inclinometers in the slurry wall were also small and amounted to no more than 3/4 inch.

After the completion of the below grade work a ventilation building was constructed on the foundations constructed with the underground shaft. Ventilation fans, silencers, electrical and other work was completed. The result of all of the work associated with this project was two new ventilation buildings and their associated equipment along with significantly improved stairs from tunnel level to street level. For a comparison of old versus new see Figures 8, 9, 1 and 2.

Figure 8. Photo of New Ventilation Building

Figure 9. Photo of Old Ventilation Building

SI Conversion

1 acre = 0.4 ha
1 acre-ft = 1,233 m^3
1 ft - 0.3 m
1 sq ft = 0.09 m^2
1 cu ft = 0.03 m^3
1 gal (U.S.) = 0.004 m^3
1 in. = 25.4 mm
1 sq in. = 645 mm^2
1 lbf = 4.5 N
1 lbm = 0.5 kg

1 psf = 48 Pa
1 psi = 6.9 kPa
1 mgd = 0.04 m^3/s
1 mi = 1.6 km
1 sq mi = 2.6 km^2
1 short ton = 907 kgf
1 yd = 0.09 m
1 sq. yd = 0.8 m^2
1 cu yd = 0.8 m^3

Building Protection from Tunneling In Downtown Los Angeles

Loring A. Wyllie, Jr.[1] and John A. Dal Pino[2]

Abstract

The Los Angeles Metro Rail Red Line involves twin 20 foot diameter bored tunnels between underground stations in downtown Los Angeles. The A146 segment includes a 90 degree curve from Hill Street to 7th Street with tunneling directly beneath a series of multi-story buildings. This paper describes the analysis of these structures with regard to tunneling effects and the building protection measures employed.

Introduction

The Los Angeles Metro Rail Red Line makes a 90 degree turn in downtown Los Angeles. Heading south out of the 5th and Hill Street Station, the A146 segment alignment curves right, taking two city blocks in both the south and west directions to head west beneath 7th Street. The alignment is shown in Figure 1. The tunnels are about 20 feet in diameter and are spaced approximately 43 feet on center. The tunnels were constructed using an open-faced tunnel boring machine (the shield) and have temporary precast concrete liners and permanent cast-in-place concrete walls.

The alignment is directly beneath a series of multi-story buildings, varying from four to fourteen stories in height. Most of the buildings are older buildings constructed between 1910 and the mid-1930s, although several were built in the 1960s and 1970s. Most have a single basement although one has a deep sub-basement mechanical room. All buildings are supported

---
[1]Senior Principal and Chairman, H. J. Degenkolb Associates, Engineers, 350 Sansome Street, #900, San Francisco, CA 94104
[2]Principal, H. J. Degenkolb Associates, Engineers, 350 Sansome Street, #900, San Francisco, CA 94104

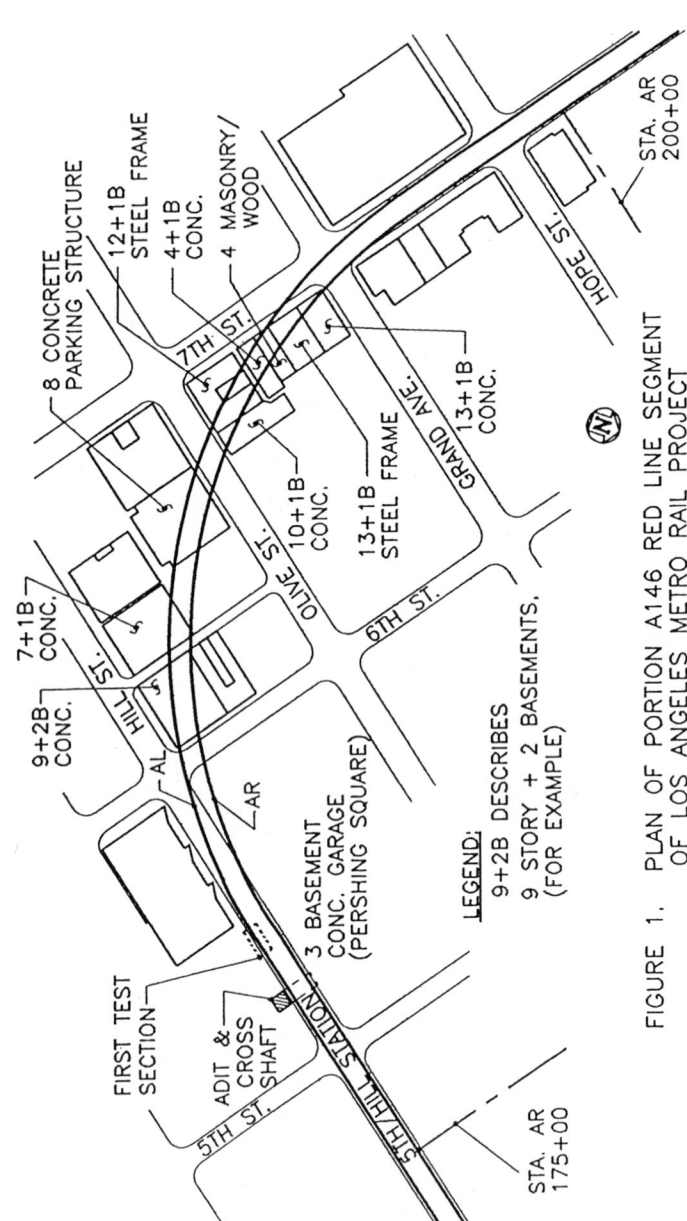

FIGURE 1. PLAN OF PORTION A146 RED LINE SEGMENT OF LOS ANGELES METRO RAIL PROJECT

on spread or combined footings, probably designed for bearing pressures in the 5 to 8 kips per square foot range.

The soils are alluvial in this section of the alignment. The preconstruction geotechnical report indicated that the soils are granular alluvium consisting of clean sands and gravels with some silty sands, gravelly sands, sandy gravels, cobbles and boulders. The alluvium is very dense at tunnel elevations and were reported as relatively cohensionless. The ground water table is well below the tunnel inverts so tunneling was in the dry and unpressured.

Estimated settlements of the structures above and adjacent to the tunnels were prepared by others and provided to us. The tunnels vary in depth below grade, with a depth about 60 feet below grade to the crown of the tunnel midway in the curve and about 40 feet below grade at both ends of the curve. This places the tunnels about 20 to 40 feet below the building footings. The estimated settlements of building footings varied from 1 inch to 3 inches depending on the depth of the tunnels with differential settlements between adjacent columns obviously much less because of the relatively flat nature of the settlement curve for the twin tunnels.

The questions to be answered and related problems to be solved were as follows:

1. What are the expected effects on the buildings above and adjacent to the tunnel alignment from the tunneling?

2. What building protection measures are appropriate to limit or minimize damage to these buildings?

3. If there is an unexpected accident, or run, in the tunneling beneath a building, what might be the worst case scenario? Are any special protection measures needed?

Analysis of the Problem

Buildings are not damaged by total settlements, they are damaged by differential settlements. A building that settles uniformly 1 to 2 inches will be undamaged except for the possibility of damage to underground utility connections or the grades at entries. But differential settlements of adjacent columns can rack a building causing damage to the structural frame, cracking in architectural finishes and unlevel floors. The effects of tunneling on a building is a complex problem, since the timing and rate of advancement of the tunneling must be considered in addition to the impact of the estimated final settlement trough on the building. Quite often, one tunnel is driven many months ahead of the second and there is also a bit of a "ripple" effect along the alignment as the tunnel shield works its way beneath a building. However, in many soils, the time dependent effects of soil movements will overshadow these "ripple" effects. There is also concern for the rare accident, or run, which causes severe settlements immediately above the location of the run or instability. In the early design phase for this segment, there was concern whether it was safe to tunnel beneath these buildings while they were occupied.

To simplify this analysis procedure, we elected to bracket the problem by addressing the worst case scenario. What would happen to a building along the alignment if there was a run directly beneath a column and bearing for the column's spread footing was completely lost? Could there be a partial collapse? Or would the building support itself with a sizeable sag in the floors and considerable damage? This procedure was conducted for all buildings directly above the tunnel alignment. Then, a scaling procedure was used to estimate the building performance responding to the estimated maximum settlements of 1 to 3 inches in the shape of a settlement trough provided to us.

The analysis procedure consisted of a limit state approach assuming complete loss of support of selected columns above the alignment. Drawings were available for all but one building along this segment of the alignment, which greatly aided the analysis. The structures were analyzed assuming the beams, slabs and walls in both directions would attempt to resist the loads of the column without support. In many cases, this came down to relying on the continuous reinforcing steel and the tensile capacity of steel beams and their connections as the floors act as a series of two-way catenaries supporting the affected column. Obviously, columns in exterior walls many stories high could generally support the affected column with virtually no settlement of the column. Interior columns are more vulnerable but in most cases we found that the structure could support the affected column without collapse. We calculated an estimated settlement in this limit state and found that this worst case differential settlement would generally vary from one inch to about 12 inches. We found only one column in one building where we could not discount a partial collapse if full bearing was lost beneath its footing.

Based on the results of this analysis, combined with the very small chance of a run, it was felt that it was safe to tunnel beneath these buildings while they were occupied. Furthermore, due to the large settlements that were predicted due to normal tunneling, it was our opinion that the buildings would be significantly cracked and damaged with extensive repairs required. Therefore, we recommended that building protection in the form of compaction grouting be employed to limit building settlement. In two cases, where tunneling was very close to adjacent footings (e.g., within 6 feet), we recommended underpinning procedures to support the building loads below the tunnels. In another case, a multi-story parking structure above grade with maximum estimated settlements of about 1 inch was recommended to have no specific protection.

Building Protection

To control movement of the soil above the tunnels, a compaction grouting system was recommended. With compaction grouting, a stiff, cementitious grout is injected at very high pressure below grade through steel pipes to a arch-shaped zone just above the tunnel crowns to replace the soil which is lost into the tunnel excavation during normal tunneling operations. In theory, the volume of ground lost into the excavation is replaced with the cement grout and void spaces are compressed, with the overlying soil supported from below resulting in minimal settlement of the surface soils. The cement grout is injected in the loosened soil zone approximately 5 feet behind the tunnel shield as the shield progresses along the alignment. Grout is injected continuously in several injection pipes after the shield passes until the ground refuses to accept additional grout. We did not inject the grout above the shield for fear of binding the shield in the ground. We hoped to inject the grout as the voids in the soil formed, and before the voids could propagate upward toward the building foundations.

The compaction grout consisted of a mixture of cement, sandy loam and water with a very stiff consistency. The stiff grout was used because a more plastic grout can fracture the ground horizontally and flow laterally and thus not generate the necessary compaction, ground displacement and support action.

The grout was injected at four points arranged on a circular arc located 10 feet radially outward from the tunnel crowns. Each tunnel has an independent set of grout points, as seen in Figure 2. Along the arc, the grout points were spaced about 10 feet apart. Along the tunnel alignment, groups of four grout points were spaced 10 feet apart but staggered 5 feet apart within an individual group to create a uniform pattern of injection points.

The 10 foot spacing between the grout bulbs and the tunnel crowns was the critical distance in the compaction grouting scheme. It was our desire to minimize the spacing between the grout bulbs and the tunnel crowns to best control ground movement, although based on our experience on previous projects, we believed that a separation of less than 10 feet would lead to excessive pressure on the erected temporary tunnel lining and shield which could lead to a blowout and soil run into the tunnel and binding of the shield. The other 10 foot spacings, laterally and along the alignment, were based on the effective zone of influence of compaction grouting.

The grout was injected in 2 to 3 inch diameter pipes that were installed much in advance of the actual tunnel construction. Actual pipe diameters were selected by the Contractor for adequate stiffness in driving to insure accurate installation. The pipes were drilled into the ground along radial lines targeted for the center of each tunnel. This pattern was selected to allow the pipes to be advanced or retracted if necessary to allow the injection of additional grout in essentially the same location. This preferred grout pipe geometry was altered as required to miss underground utilities, building foundations and other below grade obstructions.

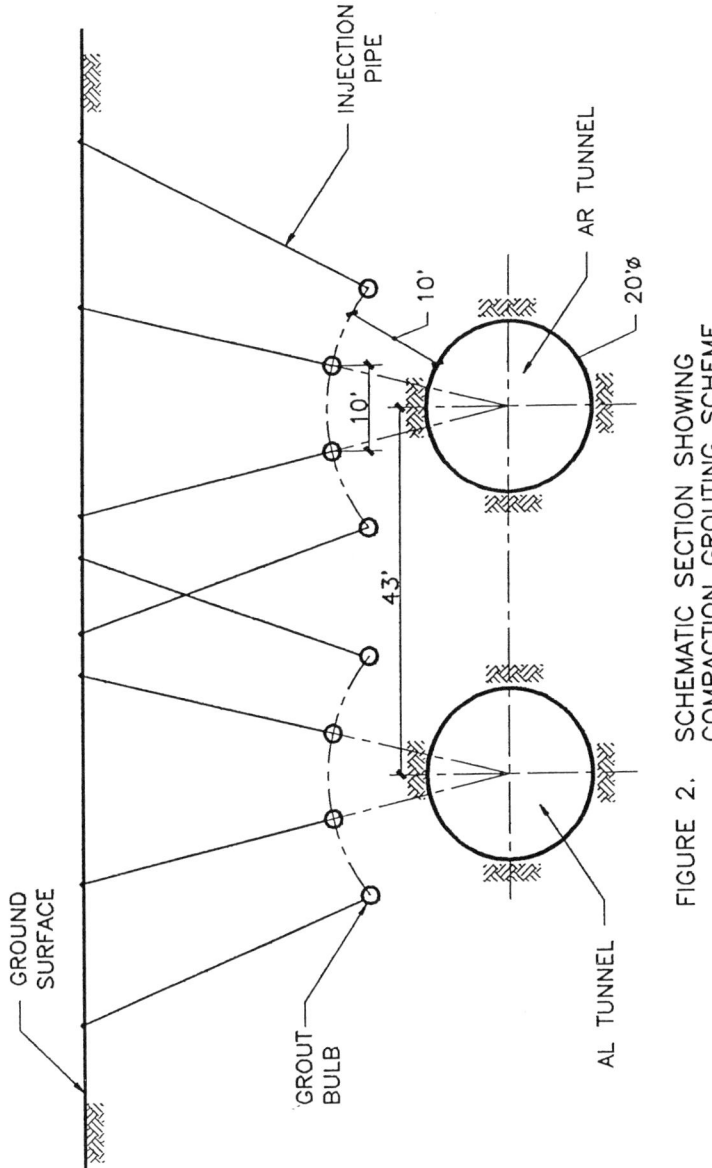

FIGURE 2. SCHEMATIC SECTION SHOWING COMPACTION GROUTING SCHEME

In order to test the effectiveness of the compaction grouting procedure, a test section was specified in the middle of Hill Street after tunneling about 100 feet of the first tunnel, AL. The test section was 50 feet long and heavily instrumented with extensometers and surface survey points to evaluate performance. The project specifications allowed for adjustment in the grouting procedures by the Tunneling Contractor based on performance within the test section. At the test section, the Tunneling Contractor determined the proper grout mix and experimented with jet-grouting to create intentional voids.

At two structures, the tunnels were so close to existing foundations that it was decided that compaction grouting would not be appropriate. At these locations, underpinning was installed. One example is the northern portion of the Pershing Square Garage. The corner of the garage was modified to accommodate the end of the 5th and Hill Street Station and the adit and cross shaft for the beginning of the AR tunnel which were about 6 feet below the foundations of the exterior wall of the garage, as seen in Figure 3. Since there was no space to use compaction grouting, underpinning was installed. Compaction grouting was resumed when the tunnel had dropped sufficiently in elevation to allow grouting. The underpinning was generally as shown in Figure 3, with pits, measuring 6 foot by 12 foot in plan, excavated in two stages by hand digging procedures. Thick, story-deep cantilever beams picked up the perimeter wall, where a story-deep wall was cast against the segmental existing exterior wall to act as a beam spanning between the cantilevers from the underpinning pits. Jacks were installed at the top of the pit below the cantilever wall to pre-load the underpinning system. These jacks were left in place during tunneling in case a soil wedge above the foundation bonded to the exterior wall and overloaded the underpinning system. Some of the cantilever beams were on curving, circular ramp walls which complicated the underpinning details.

Construction Results - Test Section

The tunnels were specified to be driven from 5th and Hill Street Station towards 7th Street with the AL tunnel driven first. This allowed the heavily instrumented test section in the middle of Hill Street to verify the compaction grouting procedures before tunneling under the first building. The soils were very extensively logged during installation of grout pipes and instrumentation in the test section. This logging

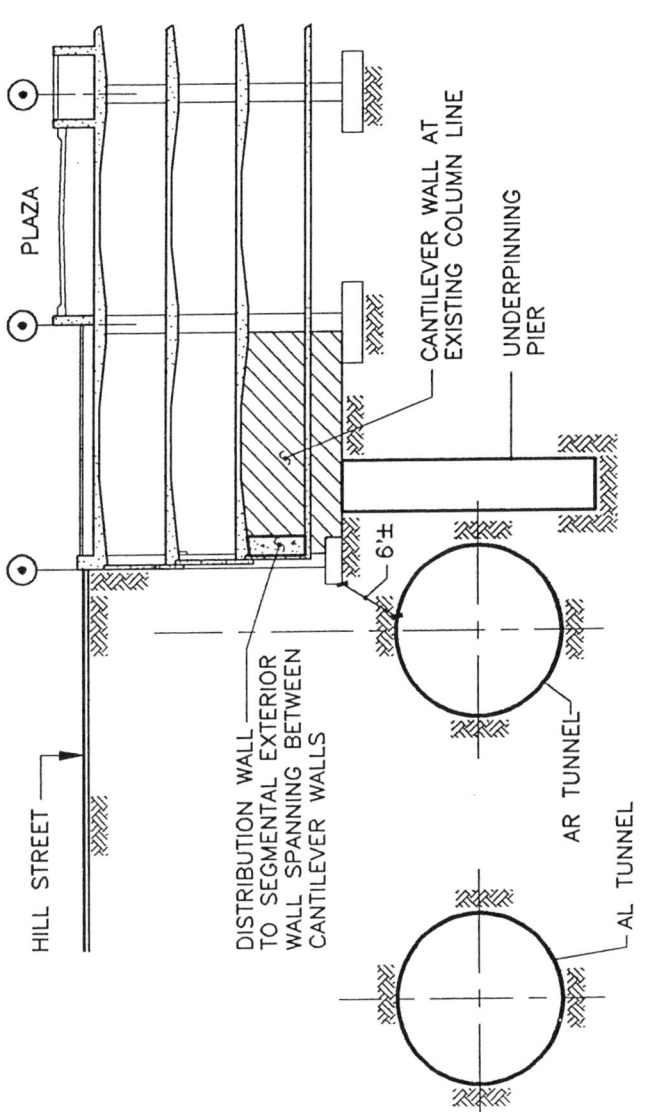

FIGURE 3. UNDERPINNING OF PERSHING SQUARE GARAGE

revealed a series of alluvial layers at the level of the tunnels and in the region where the compaction grouting was specified. Several of these layers contained considerable fines with silts and clays present in the alluvial layers. The result was that the ground generally would not accept any significant amount of the compaction grout material. However, settlements were considerably less than expected, with maximums at street level of about 3/8 inch instead of the 2+ inch prediction without compaction grouting.

There was an extensive review of the results from the test section. It was finally concluded that the fines in the alluvium were not allowing the compaction grout to perform as anticipated since the ground did not appear to loosen significantly just above the tunnel crowns. However, settlements were considerably less than predicted considering the lack of grout injection. In effect, this "solid" layer acted as a beam which spanned over the tunnel crowns, helping to control settlement.

A second test section was performed in Hill Street before tunneling proceeded beneath the first structure. The purpose of this second test section was to experiment with means for reducing the volume of lost ground such as additional jacks for expanding the precast tunnel liner segments, injecting grout through the tunnel liner during expansion to fill voids and placing grout and pea gravel into the invert below the segments to lubricate the liner system. Additional borings were also taken along the alignment to better define the alluvial layers along the tunnel route and experiments were conducted with chemical grouting. The results at the second test section were encouraging, with surface settlement of about 2/10 inch maximum.

Based on the results of second test section and anticipated favorable soil conditions beneath the first building, tunneling was directed to proceed using the improved liner expansion procedures. Since the compaction group pipes were in place and needed to be removed, grout was injected through all pipes as specified despite the small grout volumes that were expected. There was no damage sustained by the building and settlements were small and within acceptable limits.

While proceeding beneath the surface parking lot beyond the first building, a region of soil was encountered with a poorly graded, cohesionless "sugar sand" at the crown of tunnel which easily raveled. Two runs occurred in this "sugar sand" material consisting of several yards of materials, although the layers above the "sugar sand" arched over the run and surface settlements were not appreciable. The compaction grout pipes were used to fill the voids with grout.

Again, there was a period of review. Due to the uncertainty of the locations of the various alluvial layers and the fact that borings confirmed that more "sugar sand" was ahead, it was decided that it was prudent to chemically grout the soil near the crown of the tunnel to prevent runs of soil when "sugar sands" or similar sandy soils were encountered. The Tunneling Contractor performed the chemical grouting by drilling two pipes ahead about 60 feet and slightly upward from the open face through the shield and chemically grouting while the pipes were retracted. This procedure was repeated after the shield had advanced to the end of the chemically grouted soil. With this procedure, no further significant problems were encountered and the AL tunnel was completed and the AR tunnel was drilled successfully. Although the authors never received copies of the extensive surveying and building monitoring data, it is our understanding that most building settlements above the tunnel alignment were within the 1/8 to 1/4 inch range and we are not aware of any damage to any of these buildings along the curved section of the A146 segment.

Conclusion

In conclusion, the A146 segment of the Los Angeles Metro Rail Red Line in downtown Los Angeles presented some interesting challenges in terms of protecting buildings from the effects of tunneling. Settlements up to 2 to 3 inches were predicted without protection measures which could cause extensive damage to buildings.

Buildings were analyzed to determine their stability if complete loss of support occurred beneath a footing above the tunnel alignment. Based on the ability of these structures to work as a series of floor catenaries and prevent building collapse, it was decided it was safe to allow continued occupancy during tunneling. Due to the large anticipated settlements from normal tunneling, compacting grouting was recommended to fill the voids in the soil as they occur and minimize building settlement.

Two test sections revealed that the soils were better than anticipated with much smaller actual settlements than those predicted. The soil layers contained more fines than anticipated, preventing the injection of significant quantities of compaction grout. Despite this, compaction grouting was performed to the extent possible for added settlement control and building safety. Procedures were developed to improve expansion of the precast liner segments and reduce loss of ground. After two small runs under a surface parking lot due to a layer of "sugar sand" at the tunnel crown, a chemical grouting scheme of the crown through the shield was instituted which eliminated further problems on the segment. Settlements were minimal in all buildings and the authors are not aware of any settlement-induced damage in any of these buildings.

Acknowledgement

The authors wish to thank the late Henry J. Degenkolb, whose judgment and intuition guided this project to its successful conclusion. We appreciate the support and confidence of the Metro Rail Transit Consultants and Dr. James E. Monsees throughout this project. The assistance of Ed Graf relative to compaction grouting techniques was most valuable. The project also owes it successful conclusion to the support of the Owner, the Southern California Rapid Transit District; the construction manager PDCD, consultants Ralph Peck, Tor Brekke and James Gould and the good tunneling practices of the contractor, Shank-Ohbayashi.

Deep Cuts and Ground Movements in Chicago Clay

Richard J. Finno[1]

ABSTRACT

Ground movements adjacent to deep excavations made through soft Chicago clay are reviewed. Typical subsurface conditions in the downtown area of Chicago are described and undrained shear strength variations are discussed. The effects of wall and deep foundation installation, cycles of excavation and bracing, and strut removal on observed movements are illustrated through review of case histories. The importance of considering all aspects of construction which impact stresses in the ground adjacent to the supported wall when making a preconstruction estimate of magnitude and extent of ground movements is emphasized.

INTRODUCTION

Perhaps there is no other situation where construction details and geotechnical considerations are so intertwined in impacting the performance of an earth structure as for supported excavations made in soft clay. Seemingly minor construction details or geologic variations can have unproportionally large impacts on observed performance of an excavation and its surroundings. To illustrate this conundrum, ground movements adjacent to deep excavations made through soft Chicago clay are reviewed. All excavations were made in the downtown area of Chicago where the geologic conditions are similar, but where significant differences may exist. Ground movements associated with wall and deep foundation installation, cycles of excavation and bracing, and strut removal are illustrated through case studies. The relation between responses observed during one portion of construction and those observed in latter stages is noted. As has been shown before (eg. Peck, 1969; Mana and Clough, 1981) S_u is the most important geotechnical indicator of potential ground movements with excavations in clay. To

---

[1] Associate Professor, Department of Civil Engineering, Northwestern University, Evanston, IL 60208

evaluate a priori these movements, it is important to accurately define the undrained shear strength, S_u, in the strata affected by construction. Therefore it is important to understand the geological environment in the downtown Chicago area and its impact on the variations of S_u which can naturally occur.

This paper describes typical subsurface conditions in downtown Chicago, discusses undrained shear strength variations in the compressible clays which impact deep excavation response, reviews performance data which illustrates movements associated with various aspects of construction, and discusses the present state-of-the-art of movement prediction in light of these performance data.

SUBSURFACE CONDITIONS

Most of the soil in the downtown area of Chicago was deposited in relatively distinct sheets during local advances and retreats during the Wisconsinan Stage of Pleistocene glaciation (Frye and Willman, 1960; Larsen, 1973). The repetitive processes of advance and retreat were marked by terminal moraines and left readily identifiable strata. In order of deposition, the geographic names of the prominent moraines in the area include the Valparaiso, Tinley, Park Ridge, Deerfield, Blodgett and Highland Park (Otto, 1963; Larsen, 1973). Figure 1 shows a location plan of the area with some of the more recognizable geologic features. For a more detailed discussion of the local geology and its impacts on engineering properties, see Chung and Finno (1992).

Otto (1942) developed stratigraphic correlations for morainic strata in the Chicago area on the basis of water content and unconfined compressive strength variations. The softer and more compressible clays are found within the Deerfield and Blodgett strata. Laboratory evaluations of S_u indicate that these soft soils exhibited normalized behavior when proper consideration of natural water content was given (Finno and Chung, 1992).

The addition of the water content variable changes the nature of the normalized undrained strength relation, S_u divided by vertical effective consolidation stress, σ'_{vc}, as compared to more uniform deposits. The normalized strengths in compression and extension decrease with increasing water content, w, perhaps reflecting increasing plasticity index, and decreasing OCR. Finno and Chung (1992) presented laboratory results which indicated the following normalized S_u values for triaxial compression, TXC, and extension, TXE, tests for OCRs less than or equal to 4:

$$(S_u / \sigma'_{vo})_{TXC} = 0.46 \, (0.90 - w) \, OCR^{0.9} \quad (1)$$

$$(S_u / \sigma'_{vo})_{TXE} = 0.31 \, (0.90 - w) \, OCR \quad (2)$$

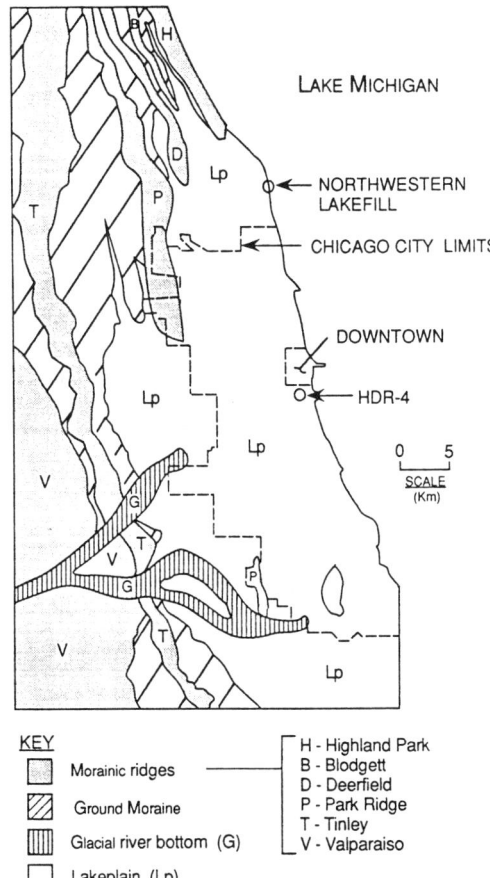

Figure 1. Location Map

Given a natural w, OCR and mode of shearing, equations 1 and 2 can be used to estimate S_u of the Blodgett and Deerfield strata.

Examples of stratigraphic and water content variations in the downtown area are given in Figure 2. Selecting values of S_u for Deerfield strata is a rather direct process because the water content generally is quite uniform. However, for genetically complicated strata like the Blodgett with its rather wide variations in water content, use of equations 1 and 2 supplements the considerable engineering judgement required to select representative values.

An example of the variation in shear strength caused by erratic water contents is shown in Figure 3 for the HDR-4 site. Using average OCR values of 1.1 and 1.2 for the Blodgett and Deerfield strata, respectively, and water content distributions shown in Figure 2, S_u values for compression and extension loadings are plotted versus depth. Ranges for compression tests are shown to illustrate possible S_u variation as a result of water content variation. The ranges represent water content variations of $\pm$ 10 and $\pm$ 3 percentage points for the Blodgett and Deerfield strata, respectively. The variability of S_u will be significantly larger in the Blodgett, indicating that locally soft layers can be expected within this stratum.

Also shown are S_u values from unconfined compression (U), unconsolidated undrained (UU) triaxial compression tests and field vane tests (Finno, et al., 1988). These results are quite scattered by a degree larger than that caused both by stress-

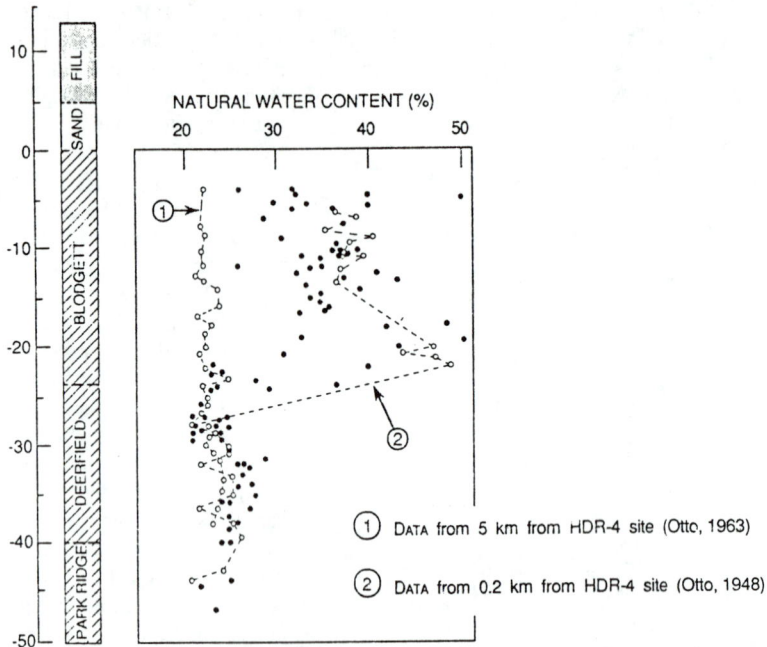

Figure 2. Typical Stratigraphy and Water Content Distributions

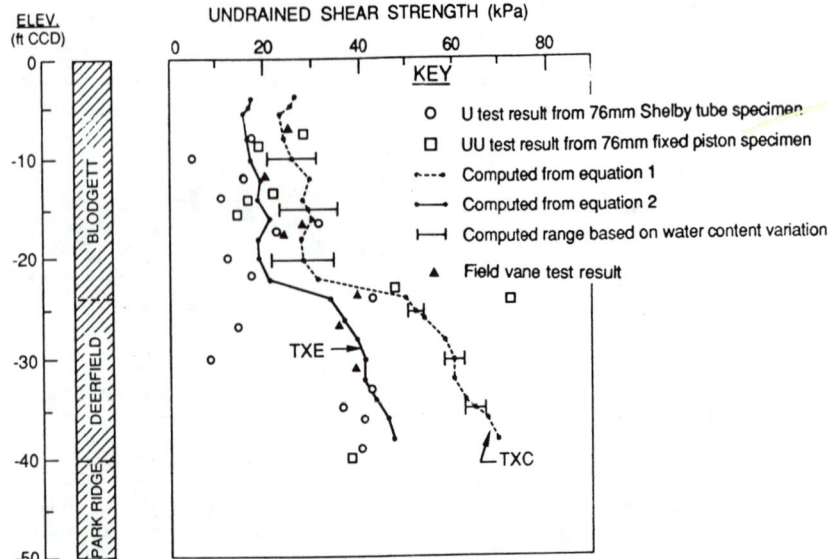

Figure 3. Typical Undrained Strength Variations for the HDR-4 Site

induced anisotropy and variable water contents, even in the Blodgett. Thus use of these strengths directly may cause one to miss important variations in S_u.

SOURCES OF GROUND MOVEMENT

Ground movements arise from many sources during construction. Those which arise from normal construction cycles of excavation and bracing can be estimated for many cases by procedures proposed by Mana and Clough (1981). However other construction activities can cause movements and/or impact the development of subsequent movements. These activities may include wall installation, pile driving, pier construction and support removal.

Wall Installation

Many excavations in Chicago were supported by sheet-pile walls and internal bracing. At the HDR-4 test section (Finno et al., 1989; Finno and Nerby, 1989), ground instrumentation was installed prior to driving the sheet-pile and therefore provides a record to directly ascertain its effects on the surrounding soil. At this site, a 40-ft-deep cut made for a subway extension located approximately 1 mile south of the downtown area (Figure 1).

Transverse soil displacements during installation of the west sheet-pile wall are presented in Fig. 4 for construction days 54 and 66. See Finno et al. (1988) for a more complete description of soil response during sheet-pile installation. These days correspond to the time when the sheet pile was installed to elevations -28 ft and -50 ft Chicago City Datum (CCD), respectively, in the vicinity of the test section. The vectors consist of the combination of extensometer data and inclinometer data obtained at elevations of the extensometer anchors. Maximum incremental displacements were approximately 0.5 inch. Upon initial driving, soil at elevations above the bottom of the penetration displaced as much as 0.4 inch. After final seating of the sheet piles, the soil at lower elevations displaced away from the sheeting, but this time with a more pronounced horizontal component.

Note that the total magnitudes of the outward movements could have been expected. Within the saturated clays, an estimate of the lateral displacement can be made by assuming that the quantity of soil displaced during driving is equal to the volume of the section. The PZ-40 section displaces an equivalent of one inch of soil per foot of section. By assuming that equal amounts displace in each direction, 0.5 inch lateral movement could be expected.

Sheet-pile driving also caused time-dependent movements of the ground surface. To show the relationship among movements pore pressures and pile driving, vertical displacements at the ground surface and the top of the clay and pore water pressures during sheet-pile installation are plotted versus time on Fig. 5. Piezometers P1-3, P2-3 and P3-3 are located 8, 14 and 28 ft, respectively, from

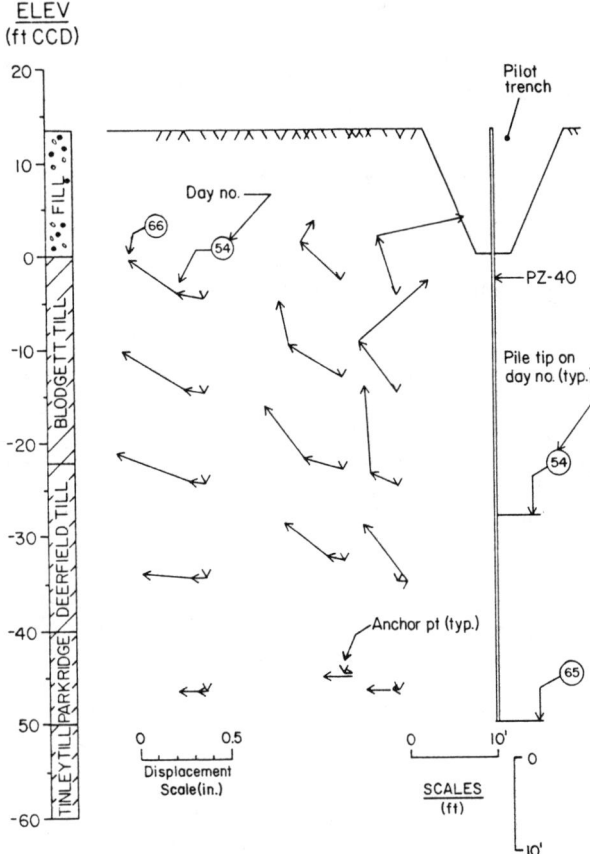

Figure 4. Ground Movements during Sheet-pile Installation at HDR-4

the west sheet-pile wall and are set at approximately elevation -31 ft CCD. Pore pressures rose rapidly in response to each pass of the sheet pile driving. The largest increases were observed after the second pass on day 64. Thereafter, the rate of pore pressure dissipation rapidly decreased with time. The ground surface, indicated by movements measured at settlement points S1, S3, and S6 (located 12, 22 and 30 ft, respectively, from the wall), and the top of the clay stratum, indicated by movements measured at extensometers E4, E2 and E3 (located 14, 21 and 40 ft, respectively from the wall) heaved when the sheet pile was installed. As pore pressures dissipated, consolidation occurred, and both the ground surface and top of the clay subsided.

Very little data exists concerning pore pressures associated with sheet-pile driving. In the few cases where pore pressures have been measured, positive pore pressures were always recorded (Finno and Nerby 1989). However, pore pressures have been recorded while driving single piles by many investigators. Fig. 6 shows the maximum ratios of excess pore pressure to effective overburden stress versus normalized distance from the sheet piles at HDR-4. Also shown is a band that reflects pore pressure response caused by driving single piles as reported by Hagerty and Garlinger (1972). Note that an equivalent diameter of 1.34 ft, equal to the depth of the PZ-40 section was used to compute normalized distance.

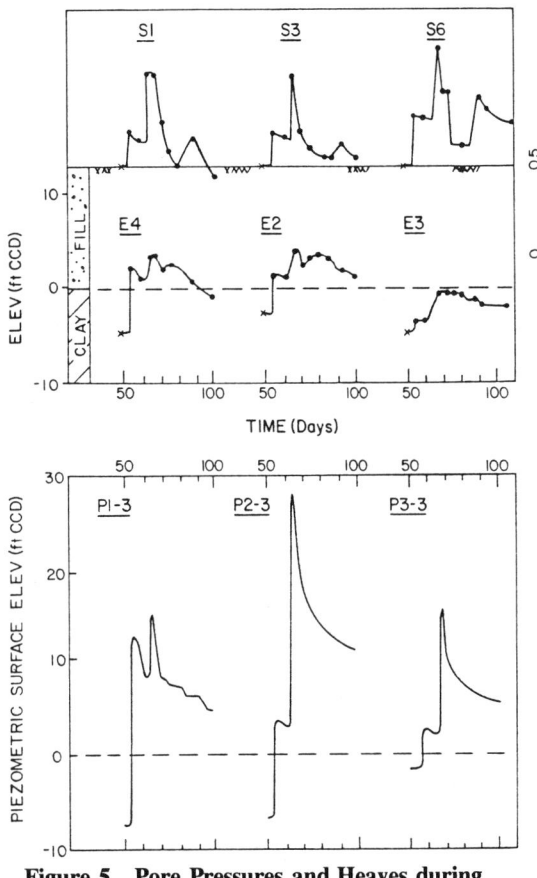

Figure 5. Pore Pressures and Heaves during Sheet-pile Installation at HDR-4

Data on Fig. 6 indicate that the extent of excess pore pressures can be larger when driving sheet piles than when installing piles. The pore pressures adjacent to the sheet piles could not be measured because of the presence of the pilot trench; however, the magnitudes are expected to follow the same trends as piles. The wider zone of influence for the sheet piles can perhaps be attributed to the differences caused by the plane strain conditions associated with sheet piles in contrast to the axisymmetric conditions associated with piles. In the same way that pile driving is known to change the stress conditions around a pile and impact its unit side resistance, one must expect differences from the in-situ stresses and pore pressures adjacent to sheet piles after a wall is installed. These effects must be recognized and considered explicitly when evaluating expected movements, otherwise any prediction will either be in error to some degree or a result of compensating errors.

Installation of Deep Foundations

Driven Piles

Driven piles were used in conjunction with several deep excavations in Chicago in the 1950s. For instance, Peck and Berman (1961) described the performance of a multi-level excavation made for a hospital in the north part of the downtown area. The deepest portion of the excavation extended 56 ft below the

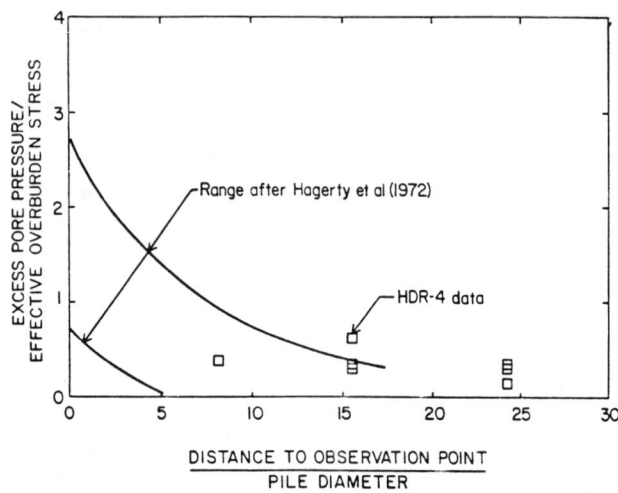

Figure 6. Sheet-pile and Pile Induced Pore Pressures

ground surface. Excavation for the various levels was carried to final grade with soil support provided by sheet piles and cross lot bracing where needed. Only then were the displacement type piles driven. Not only did the ground laterally displace and the piles heave, but bracing system displaced laterally and the bottom of the deepest excavation heaved significantly. Preaugering was tried unsuccessfully to minimize the heave caused by the pile driving. Deep foundation support was finally provided by substituting hand-excavated caissons for several hundred remaining piles.

A similar situation was described by both Peck and Berman (1961) and Hagerty and Peck (1971). More than 400 Raymond step tapered piles with 9-3/8 inch tip diameters were driven to depths of 40 ft within clusters with 2.5 ft spacings. Most of the excavation was just 5 ft below grade, but an approximately 20 ft deep elevator pit was excavated into soft clay. Movements as a result of pile heave were exacerbated by excavation of the pit. Most of the piles were judged unacceptable because of the excessive vertical and lateral movements they experienced. Costly remedial measures included driving additional H-piles to adjust the center of gravity of pile groups, tying together several groups with reinforced concrete grade beams, and conducting extensive proof load tests.

The situation was substantially better at the Inland Steel Building constructed in 1956. A soldier pile and lagging wall was installed around the deep excavation and the grade was lowered 9 ft. Low displacement H-piles were driven from this level above the final cutoff grade. While no significant problems arose from pile heave, it was difficult to precisely locate the piles at the cutoff elevation, especially for the follower-driven batter piles, leading to longer construction times for the deep foundations.

The experience gained at several sites emphasized the classic dilemma of construction sequencing under these circumstances. If piles are driven first they will be subjected to lateral movements associated with stress relief from excavation unloading. Excavation around the piles is time-consuming and therefore costly. Additionally, because piles are not driven absolutely vertically, the positions of the piles beneath the deeper parts of an excavation may vary considerably from those called for in the plans. Alternatively, if the excavation is made first, subsequent pile driving through saturated clays will produce heave, lateral movements and increased loads in the lateral support elements.

In any case, an estimate of the magnitudes of vertical ground movements associated with pile driving can be made using a procedure described by Hagerty and Peck (1971). However the stress and pore water pressure changes associated with pile driving alter the stresses inside an excavation. The available passive resistance afforded by the soil between two walls will be affected by these changes. The impact of these stress changes on subsequent support system response to excavation and bracing should be similar, but perhaps more dramatic, than that for sheet-pile effects, which will subsequently be discussed.

Drilled piers

The problems associated with using driven piles in conjunction with deep excavations and the existence of relatively sound dolomite at not excessive depths, led to the use of high capacity drilled piers, or caissons, as the preferred method of transferring load to deep competent strata in Chicago. The idea of preaugering a deep foundation to minimize ground movements in saturated clays is well accepted. However, significant movements can arise from various sources during construction of a large diameter drilled pier, primarily from squeeze while drilling through soft clays and from dewatering rock caissons.

Ground loss from squeeze:

The effects of caisson construction on ground movements is illustrated herein with the performance of the excavation and foundation for the CNA building (Cunningham and Fernandez, 1972; Baker and Gill, 1985). Figure 7 shows the lateral movements and surface settlements adjacent to two sides of the CNA excavation. Lateral movements were obtained at inclinometer locations immediately behind the slurry wall when caisson construction was complete and when excavation was complete. Inclinometers were installed and initialized after the slurry wall was constructed, but before any significant caisson construction. Settlements behind the wall represent those at the end of excavation.

The inclinometer deflections marked 1 in the figure represent the lateral movements associated with caisson construction. These movements reached as much as 2 inches along the west wall and 3/4 inch along the south wall. The movements extended well below the bottom of the excavation at both locations and were mostly restricted to the soft to medium clay. Subsequent movements were also approximately twice as large along the west wall, even though construction during excavation at each wall was similar. Ground surface settlements behind the wall reached as high as 5 inches along the west wall, more than two times larger than those along the south wall.

The large differences in lateral movements were attributed by Cunningham and Fernandez (1972) and Baker and Gill (1985) to differences in caisson construction techniques. At the start of the project, the caissons were installed from a working grade of elevation 7 ft (CCD) by inserting two temporary casings, to elevation -23 ft and -35 ft CCD, respectively. These two casings were installed by augering an oversize hole, 6 inch larger diameter than the casings, and setting the casing into the hole. The annulus was not grouted at this time. The final distance to rock was augered and the permanent core barrel was screwed into the rock to obtain a water tight seal. After inspection and concreting, the annulus between the core barrel and rock was grouted as the intermediate casing was extracted. The upper casing was filled with rubble and sand and left in place to be removed as the excavation was extended.

The work started near the west wall where the large movements were observed. The squeeze was so large that when drilling an 8-ft-diameter caisson 30 ft from the west wall after the top temporary casing was set, the hole was completely shut by inward movement of the soft clay. These squeezing difficulties prompted a change in construction procedure whereby the temporary casings were installed under a head of site clay and water to help stabilize the hole. When the hard clay was reached, the mud in the casing was removed and excavation proceeded in the dry until rock was encountered at elevation -90 ft CCD. Bentonite mud was added to balance the hydrostatic pressure in the rock as the core barrel was screwed in. Movements at the south wall during caisson construction reflected these changed procedures. The presence of the slurry in the hole provided enough support to prevent collapse of the soft clay during excavation and reduced caisson-induced movements by a factor of two.

The larger subsequent movements along the west wall were likely caused by the smaller available passive resistance afforded by the clays inside the slurry wall. The larger strains induced during caisson construction would have softened the clays inside the west wall as compared to those near the south wall. The available passive resistance near the west wall would thus be smaller and subsequent movements during excavation would be larger than along the south wall for similar excavation sequences. The importance of the passive response inside an excavation was shown for the HDR-4 excavation (Finno and Harahap, 1991).

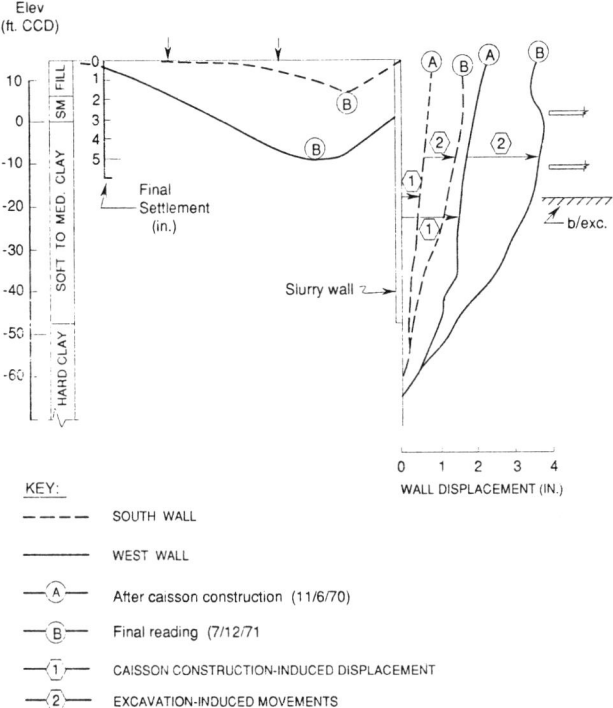

Figure 7. Wall Movements and Settlements During Caisson Installation (after Cunningham and Fernandez, 1972)

The extent of the ground surface settlements outside the excavation was two times the depth of the cut at the south wall and approximately three times the depth of the cut at the west wall (Figure 7). The extent of settlements along the south wall is typical of excavations in soft clays (Clough and O'Rourke, 1990) whereas those along the west wall are much larger than expected. This unexpected behavior resulted because of the deep-seated lateral movements which occurred during caisson construction. There are elevated train tracks supported by a steel frame located adjacent to the west wall. The large movements which occurred necessitated releveling the tracks when the differential settlements between the frame's columns reached 3 inches (Maynard and O'Rourke, 1977).

Ground movements associated with the squeeze of clays at eight sites in Chicago have been summarized by Lukas and Baker (1978). They found that squeeze will occur faster and the ground loss will be greater when the ratio of effective overburden pressure to undrained shear strength is higher. The key parameter in a priori evaluation of the potential of large ground movements when constructing caissons is therefore undrained shear strength. Estimates of movements would also depend on the diameter of the caisson, as indicated by Lukas and Baker (1978).

Ground loss from pumping:

Another source of ground movements during caisson installation occurs when shafts are extended to rock and water enters the bottom of the shaft through granular soil above the rock, seams in the rock, or from leakage along the sides of the casing. Occasionally this water contains fines which results in ground loss. These losses are largely unquantifiable prior to construction although they can be minimized with proper construction techniques.

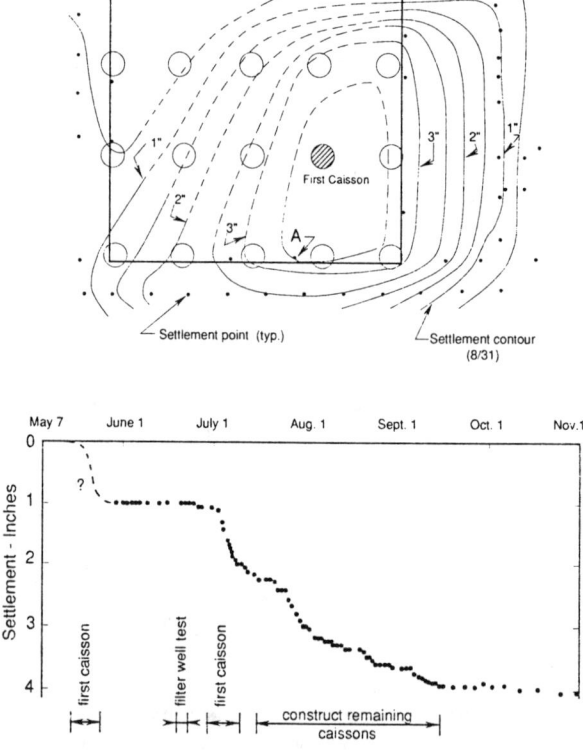

Figure 8. Settlements During Caisson Dewatering
(after Peck and Berman, 1961)

Peck and Berman (1961) describe a case where 15 hand-excavated caissons were excavated to rock. A foundation plan is shown in Figure 8. The soil just above rock consisted of 20 ft of granular soils with a piezometric level located 75 ft above the top of rock. When the first caisson was excavated to the top of the granular stratum, water was removed by pumping 500 gpm for several days. During this time an adjacent building founded on piles in tipped in hard clay above the granular stratum settled about 1 inch. The caisson was abandoned and water wells drilled to lower the water table in the sands. When a pump test was conducted, no settlements occurred at the surrounding property as indicated at point A on the figure. This observation showed that lowering the water table did not produce the settlements as a result of consolidation of the overlying soft clays, but that the settlements were a result of removing fine grained material from the predominantly granular stratum above the rock. The contractor then opened several caissons and used them as pumping wells while completing the other caissons. The settlements resulting from all these operations are also shown on Figure 8 and indicate that after the first caisson was completed, subsequent operations resulted in a total of 2 additional inches of settlement at point A.

With the advent of the technique of using a core barrel to seal off the rock from overlying granular soils, the magnitude of this component of lost ground was reduced. However, movements can still occur as indicated by performance during construction of the Sears Tower foundation (Lukas and Baker, 1978). The degree of lost ground can be seen in Figure 9 which shows a plan view near one property line and a section through this area. Table 1 shows construction details of the six caissons shown in Figure 9. The pumping rates shown in this table are significantly less than those reported in the previous case, but are still large enough to cause lateral displacements in the dense silt stratum above the rock. Approximately 4 inches of lateral movements were observed in an inclinometer within 3 ft of caisson 1 in the figure. More significantly, approximately 3/4 inch of movement towards the caissons were measured in the inclinometer at the property line, some 50 ft away.

Table 1. Caisson Construction Details (after Lukas and Baker)

Date	Caisson No.	Pumping rate (max.) (gpm)	Duration (days)
9-10 to 9-24	2	20-80	5
9-14 to 9-24	4	15-100+	5
9-21 to 9-29	6	80	2
10-1 to 10-9	1	18	3
10-2 to 10-20	3	80	1
11-20 to 11-9	5	70	5

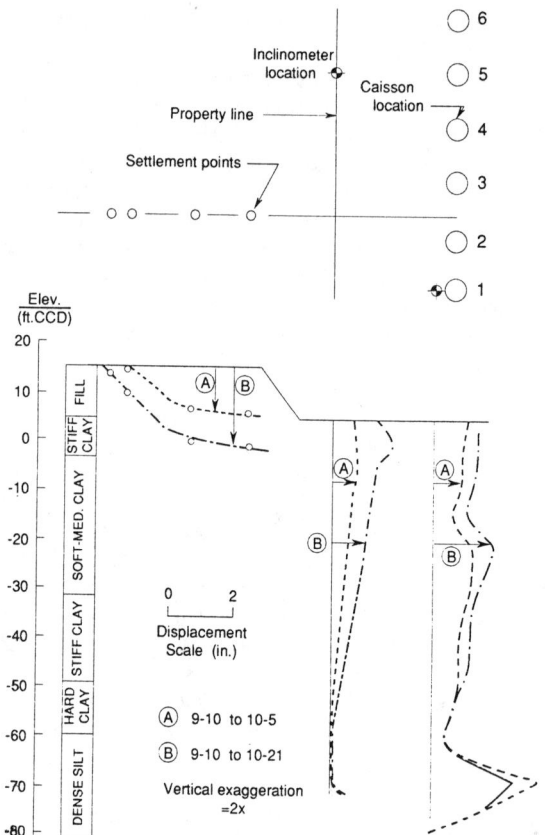

Figure 9. Wall and Soil Movements During Caisson Construction (after Lukas and Baker, 1978)

However it is somewhat difficult to separate the effects of squeeze, as indicated by the lateral movements of both inclinometers in the upper clays, and of lost ground due to piping on the settlements adjacent to the caissons. As in the CNA case, squeeze during caisson construction extended throughout the relatively compressible clays. Because the lateral movements in the soft clay had been reduced to approximately 1 inch at the property line, if the piping had no effect on the ground surface settlement, then the settlement would likely not have been as much as 2 inches by October 21. A significant portion of the settlement can be attributed to the lost ground from piping of fines into the rock caisson.

Excavation and Bracing Cycles

Wall-deformations and ground surface settlements associated with excavations in clay have been the subject of extensive study since Peck's state of the art review of the subject in 1969. Mana and Clough developed a finite element-based procedure to estimate lateral wall movements and ground surface settlements which incorporated the site geology, S_u, soil modulus, proximity of a firm stratum to the bottom of an excavation, and several aspects of construction, including preloading, wall stiffness and support spacing and stiffness. They showed that the factor of safety against basal heave is related to the lateral wall movements normalized by the excavated depth and that field data from cases with no unusual construction effects fall within a fairly narrow band, as indicated in Figure 10.

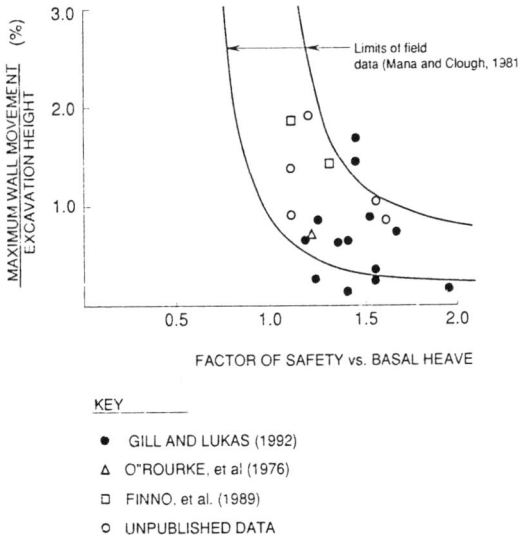

Figure 10. **Wall Movements versus Factor of Safety against Basal Heave**

Normalized wall movements for excavations in Chicago are also shown on Figure 10 and generally follow the same general trend as reported by Mana and Clough (1981). Most of the data comes from cases where the factor of safety against basal heave is less than 1.5. In these situations of relatively low factor of safety, Clough and Hansen (1981) indicated that stress-induced anisotropy should be considered in the selection of S_u when analyzing basal heave. Of the cases shown in the figure, only the HDR-4 data is based on anisotropic S_u. Note that equations 1 and 2 indicate that for low OCR, the ratio of extension to compression strengths is approximately 67 percent. Clough and Hansen (1981) indicate that unless the extension strength is less than 80 percent of its compression value, the basal heave factor of safety will be within 10 percent of its isotropic value. Therefore relatively minor changes can be expected in the other reported factors of safety in Figure 10 if anisotropy was explicitly included in an basal analysis.

The scatter in the data thus likely represents the effects of construction details. For a given factor of safety less than 1.5, the normalized wall movements from the Chicago field data varies by as much as $\pm$ 0.8 percent. For a 40 ft deep excavation, this would correspond to a variation of $\pm$ 3.8 inches, an amount which is consistent with the movements associated with construction effects, such as the caisson installation described earlier.

While the effects of construction sequencing during excavation and bracing is implicitly included in the Mana and Clough approach by specifying strut spacing which affects wall stiffness, it is necessary to consider an average value of strut spacing throughout the excavation process. Goldberg et al. (1976) also considered the effects of wall spacing on movements by plotting contours of lateral displacement on a graph of normalized wall movements versus wall stiffness. A similar approach is taken herein to illustrate the importance of construction sequencing; support spacing and therefore wall stiffness is directly affected by the amount of excavation which occurs below an in-place support before the next level is placed.

The effects of construction sequence on lateral wall movements is illustrated in Figure 11, a plot of stability number, N, (total vertical stress at the depth of excavation divided by S_u) versus normalized wall stiffness per length of wall for three excavations where detailed case records were available. This wall stiffness is the bending stiffness of the wall, EI, divided by the fourth power of the vertical distance from the lowest-placed support to the excavated surface, h, and normalized by the unit weight of water. Also shown is the maximum wall displacements which occurred from the time the support was placed until the next level of support was eventually placed. This maximum incremental movement occurred at or just below the excavated surface. The amount of excavation, h, has a large impact (to the fourth power) on the bending stiffness of the wall and, as a result, the incremental wall displacements. The largest incremental displacements occur when N is largest and wall stiffness is lowest. Conversely, there will still be some movements even when N is small and the wall stiffness is large. As suggested by Mana and Clough (1981), increasing the wall stiffness affects wall movements more when the basal heave factor of safety is low, and thus N is high.

There is dramatic increase in movements when the wall stiffness decreases and when N is large. These large incremental wall movements are accompanied by even larger ground surface settlements. The causes of the larger settlements were shown by Finno et al (1989) and are illustrated in Figure 12. This figure shows the transverse displacements which occurred when the contractor excavated below the first strut level at the HDR-4 test section. Large incremental

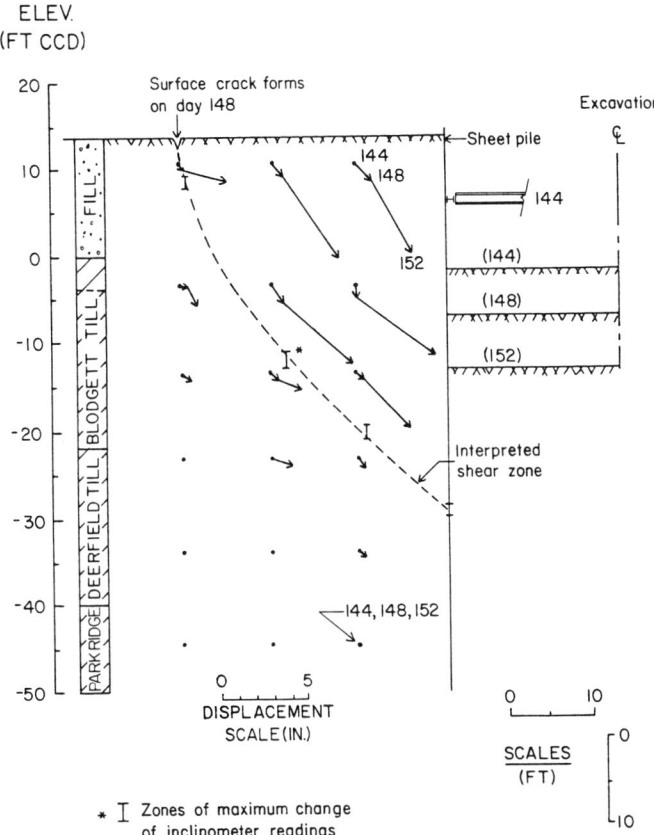

Figure 12. Incremental Displacements During Shear Band Formation at the HDR-4 Test Section (after Finno et al. 1989)

displacements occurred between day 143 and 152, when the unsupported span of the sheet-pile wall was as much as 19.5 ft over a horizontal length of 80 ft. These incremental movements are indicative of block-type movements associated with the development of a shearing surface. On day 148, a crack was observed on the ground surface indicating that a zone of concentrated strains had developed in the soil. Based on strain contours computed in the soil mass (Finno and Nerby 1989), the maximum engineering shear strain (peak value defined by Mohr's circle of strain) when the crack formed was 1.1 percent. At this time, the maximum horizontal wall movement was 2.1 inches and maximum ground surface settlement was 2.2 inches. By day 152, the maximum engineering shear strain had reached 6.7 percent, while the wall had now moved 7 inches and the ground settled 7.6

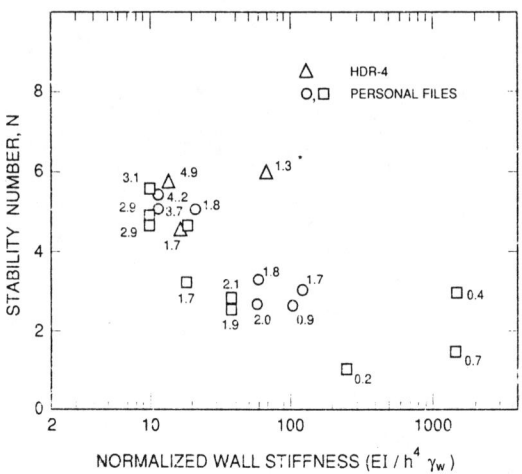

Figure 11. Incremental Wall Movements as a Function of Stability Number and Wall Stiffness

inches. The lateral wall movements will be less than vertical and lateral soil movements at some distance behind the wall because these latter movements develop in response to the unrestricted relative displacements across the shear zone, whereas the wall's continuity restricts relative soil movements.

The computed strains at this point of loss of control of ground movements agree qualitatively with those from laboratory data. The peak shear stresses in K_o consolidated, undrained triaxial compression tests occur between 0.5 and 1.6 maximum engineering shear strain (Finno and Chung, 1992). Thus, control of ground movements was lost when strains corresponding to peak shear stresses were exceeded; strain-softening and localization of strains resulted, which further promoted ground movements and caused large increases in strut loads (Finno et al. 1989).

Brace Removal

Most reports of observed movements associated with supported excavations in Chicago end with data obtained when an excavation has bottomed out. Economics and ease of data collection are the main reasons this situation is common. However removal of support can result in significant lateral movements and ground surface settlements.

For example, removal of the lower two support levels at the HDR-4 test section resulted in 0.3 inches of additional inward movements at the sheet-pile wall. This corresponded to about 7 percent of the total maximum lateral wall movement. This relatively small amount can be attributed to the reinforced concrete slab that was placed across the bottom of the excavation and the fact that the upper two brace levels were still in place when the final sets of readings were obtained. Lateral wall movements due to support removal at two sites reported by Gill and Lukas (1991) were 0.2 and 1.25 inches which corresponded to 10 and 16 percent of the final lateral wall movements.

PREDICTING MAGNITUDE AND EXTENT OF GROUND MOVEMENTS

Effects of wall installation, cycles of excavation and bracing and brace removal on ground movements may be quantified through continuum analysis, such as finite element analysis if the exact construction procedure is known, i.e. in addition to predicting soil response, one must also predict contractor and subcontractors behaviors. Even if one can see in the future to do so, effects of pile driving of deep foundations and pier construction, because of their three-dimensional nature within an excavation, are still beyond the state of the practice. But even for excavations without a deep foundation, there are limits to the state of the art. The little data that exists concerning sheet-pile effects indicate that those observed responses occur in all cases where sheet pile is installed through saturated, soft clay. The localization of strains would only occur when relatively large movements occur. Both aspects of response substantially impact the accuracy of finite element-based predictions, as subsequently discussed.

Sheet-pile Installation Effects

The observed pore pressure responses at the HDR-4 test section (Figure 5) imply that stress conditions just before excavation at a braced cut can be substantially different from K_o conditions. Stress paths were postulated based on observed pore pressures during and after driving (Finno and Nerby, 1989). These postulated stress changes are supported by results of finite element analyses (Finno and Harahap, 1991) which indicate that sheet-pile installation reduces shear stresses at approximately constant mean normal effective stress. The mobilized shear resistance on the active side of the wall is thus greater than when no sheet-pile effects are considered. Conversely, the mobilized shear resistance on the passive side is considerably smaller when sheet-pile effects are considered. Similar effects would be caused by movements associated with caisson installation.

Fig. 13 presents the results of a finite element parametric study which considers the effects of sheet-pile installation on lateral wall deformations. Variables in the study include the height to width ratio (H/B) of the excavation,

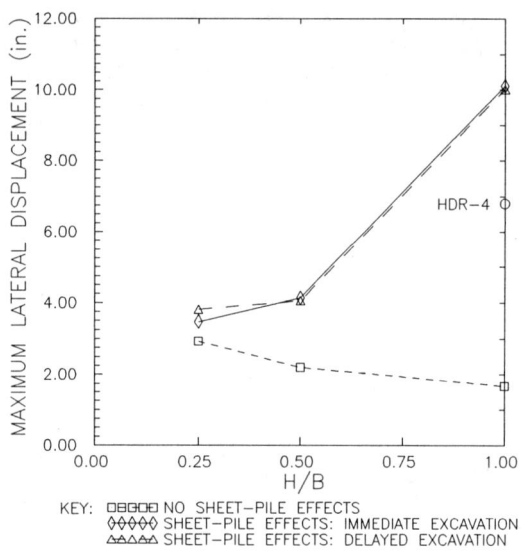

Fig. 13 Parametric Study of Sheet-pile Effects

the time between installing both sheet-pile walls, and the time between end of wall installation and beginning of installation. For a complete description of the finite element procedures used in this study, see Finno and Harahap (1991). Two limits are defined in Fig. 13. The upper limit occurs when both sheet pile walls are installed at the same time, while the lower limit results from no sheet-pile installation. The trends of this lower limit agree with those published studies of geometry effects (Mana and Clough 1981) which indicate that, all other things being equal, narrower excavations result in smaller sheet-pile movements than wider ones. When sheet-pile effects are included, this trend is reversed. Also shown are the cases when 60 days lapsed after installing both walls at the same time and start of excavation for various H/B, and the sequence for the HDR-4 test section where 60 days lapsed between installing each wall. The effects of consolidation time between wall insertion and excavation on sheet-pile displacements is seen to be minor.

Calculated results in Fig. 13 indicate that the sheet pile effects are greatest for narrow excavation. For wider excavations, the effects become smaller as the magnitude of sheet-pile induced excess pore pressures become smaller. As the time between installing each wall becomes larger, the sheet pile effects on the computed displacements become smaller. This trend is primarily a result of the smaller excess pore pressures generated in this case as compared to the situation when the walls are installed together. Walls in practice are not generally installed together so this latter condition is more representative of field situations. In any case, large computed differences arise depending on whether or not the wall installation is explicitly modeled. For situations where it is not considered, as is common in finite element analyses, any "correct" prediction of movement must therefore be a result of compensating errors.

Strain Localization Effects

The impact of strain localization on computed wall and ground displacements is shown in Fig. 14, a plot of observed versus finite-element computed movements at days 144 and 152, when the contractor was excavating below the first strut level and large movements occurred. See Finno and Harahap (1991) for a more detailed discussion of finite element computed and observed deformations. It was during this period that computed deformations deviated from

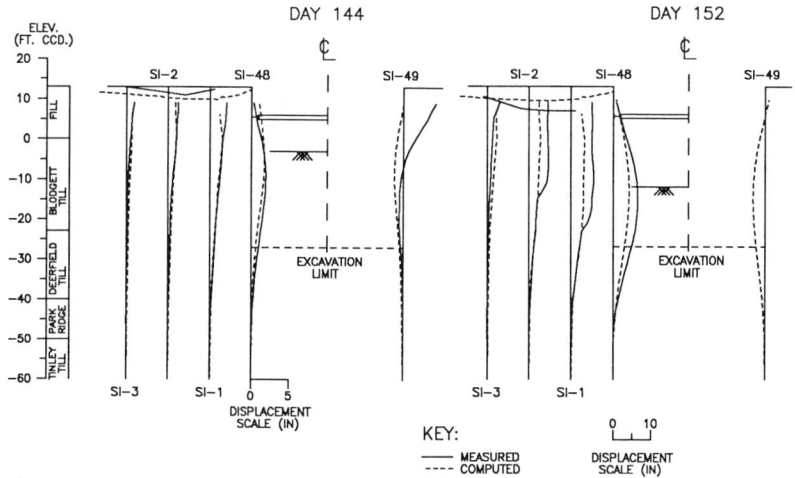

Figure 14. Computed and Observed Wall Deformations at HDR-4

those observed. After preloading the first strut on day 144, asymmetric computed and observed displacements of the sheet-pile walls are clearly seen. On this day, the typical form of sheet-pile movements is clear for both computed and observed responses. There is a rotation about the top strut, and maximum movements occur below the excavated surface. Soil movements behind the wall had the subdued deflected shape of the sheet pile. In general, computed horizontal deformations behind the wall and ground surface settlements were slightly larger than those measured. After an additional 9 ft of soil had been excavated, large lateral deformations occurred between day 144 and 152 during which maximum computed sheet-pile deflection increased from 1.7 to 5.5 inches. Measured sheet-pile deformations are slightly larger than computed deflections. Observed and computed ground surface settlements also begin to deviate at this time, with computed values being smaller closer to the sheet pile and larger at greater distances from the sheet pile than those observed. In the soil mass behind the sheet pile, the difference in computed and observed horizontal deformations is due to the increasingly wide displacement discontinuity at the locations first observed on day 148. Below this discontinuity, computed deformations are close to those observed. Thus it is at this point that the detailed trends in the soil deformations could not be tracked by the finite element simulation.

A comparison of the computed and observed strain contours (Finno and Harahap, 1991) indicates that the development of the incipient shear surfaces in the field was not captured in the finite element simulation. This inability to model localization of strains has a limited impact on the ability of the model to compute sheet-pile deformations, most likely as a result of the compatibility constraint imposed by the sheet pile. However, this drawback does prevent a reasonable estimate of ground surface settlements to be made after a certain level of strain is attained in the soil.

CONCLUDING REMARKS

Past performance data, as compiled by Peck (1969), Goldberg et al. (1976), and Clough and O'Rourke (1990) can be used to estimate expected ground movements assuming no unusual conditions (i.e. "average workmanship") develop during construction, but can not be used to estimate detailed performance. If detailed performance data are collected during excavation for the purpose of controlling movements in a project where adjacent structures are especially sensitive to movements, then this type of prediction will not yield enough information to provide a rational basis for controlling movements during excavation. In this case, a finite element-based prediction is needed.

However, the comparisons presented herein indicate that currently-used finite element simulations of braced excavation can not compute all aspects of performance. To successfully model performance, all aspects of construction which affect ground movements must be explicitly considered. For example, sheet-pile installation has been shown to have a large impact on sheet-pile deformations under certain circumstances, and yet it is usually not considered in a finite element analyses. If deep foundations will be installed within an excavation, then its effects on subsequent movements must be considered. Even if all aspects of construction are explicitly considered, a further limitation exists which is related to cases where relatively large movements arise. Practically speaking, one can expect this situation when the factor of safety against basal heave for an excavation approaches one. Then the question of shear strength selection becomes important. The accuracy of surface settlement profile is limited in these cases by the inability of commonly-used soil models to properly account for strain localization. For cases where the movements are small, for example within Peck's zone I (Peck, 1969), the finite element-based predictions can yield sufficiently accurate results.

Fortunately, computed sheet-pile deformations are not as drastically affected by these limitations and reasonable estimates of these deformations can be made even when large movements develop (Finno and Harahap 1991). In these cases, ground surface settlements are best made empirically by estimating the maximum settlement to be at least equal to the maximum horizontal sheet-pile deformation and using a distribution of settlement based on past performance data. For cases where the ground water table is not significantly altered during excavation, the maximum surface settlement is approximately equal to the maximum horizontal displacement of the soil adjacent to the wall, but not to the displacement of the wall itself. The HDR-4 field data indicated that the ratio of maximum settlement to maximum wall displacement was 1.4. Note that finite element parametric studies (Finno and Harahap, 1991) indicated that dissipation of the sheet-pile induced excess pore pressures did not significantly contribute to this ratio.

APPENDIX I. REFERENCES

Baker, C.N., and Gill, S.A., 1985, "Construction of Drilled Shafts," The Practice of Foundation Engineering, R.J. Krizek, C.H. Dowding, and F. Somogyi, eds., Northwestern University, Evanston, IL, pp. 339-369.

Chung C.-K., and Finno, R.J., 1992, "Influence of Depositional Processes on the Geotechnical Parameters of Chicago Glacial Clay," to be published in Engineering Geology, Elsevier.

Clough, G.W., and Hansen, L.A., 1981, "Clay Anisotropy and Braced Wall Behavior," Journal of the Geotechnical Engineering Division, ASCE, Vol. 107, July, pp. 893-914.

Clough, G.W., and O'Rourke, T.D., 1990, "Construction Induced Movements of Insitu Walls," Proceedings, Design and Performance of Earth Retaining Structures, ASCE Special Geotechnical Publication No. 25, ed. P.C. Lambe and L.A. Hansen, pp. 439-470.

Cunningham, J.A., and Fernandez, J.I., 1972, "Performance of Two Slurry Wall Systems in Chicago," Proceedings of the Specialty Conference of Performance of Earth and Earth-Supported Structures, ASCE, Purdue University, Lafayette, IN, Vol. I, Part 2, pp. 1425-1449.

Finno, R.J., Atmatzidis, D.K., and Nerby, S.M., 1988, "Ground Response to Sheet-Pile Installation in Clay," Proceedings of the Second International Conference of Case Histories in Geotechnical Engineering, University of Missouri-Rolla, St. Louis, MO, pp. 121-126.

Finno, R.J., Atmatzidis, D.K., and Perkins, S.B., 1989, "Observed Performance of a Deep Excavation in Clay," Journal of Geotechnical Engineering, ASCE, Vol. 115, No. 8, pp. 1045-1064.

Finno, R.J. and Chung, C.-K., 1992, "Stress-Strain-Strength Responses of Compressible Chicago Glacial Clay," to be published in the Journal of Geotechnical Engineering, ASCE, Vol. 118, No. 10, Oct.

Finno, R.J. and Harahap, I.S., 1991, "Finite Element Analyses of the HDR-4 Excavation," Journal of Geotechnical Engineering, ASCE, Vol. 117, No 10, pp. 1590-1609.

Finno, R.J. and Nerby, S.M., 1989, "Saturated Clay Response During Braced Cut Construction," Journal of Geotechnical Engineering, ASCE, Vol. 115, No. 8, pp. 1065-1084.

Frye, J.C., and Willman, H.B., 1960, "Classification of the Wisconsinan Stage in the Lake Michigan Glacial Lake," Illinois State Geologic Survey. Circular 285. Urbana, IL., 16 pp.

Goldberg, D.T., Jaworski, W.E., and Gordon, M.D., 1976, "Lateral Support Systems and Underpinning, Vol. II, Design Fundamentals," Federal Highway Administration, Report No. FHWA-RD-75-129, 249 pp.

Hagerty, D.J. and Garlinger, J.E., 1972, "Consolidation Effects Around Driven Piles," Proceedings of the ASCE Specialty Conference of Performance of Earth and Earth-Supported Structures, Vol. I, Part II, Purdue University, pp. 1207-1222.

Larsen, J.I., 1973, "Geology for Planning in Lake County, Illinois," Illinois State Geologic Survey, Circular 481, Urbana, IL., 43 pp.

Lukas, R.G., and Baker, C.N., 1978, "Ground Movement Associated with Drilled Pier Installations," ASCE Annual Convention, Pittsburgh, PA.

Mana, A.I. and Clough, G.W., 1981, "Prediction of Movements for Braced Cuts in Clay," Journal of the Geotechnical Engineering Division, ASCE, Vol. 107, No. 6, pp. 759-778.

Maynard, T., and O'Rourke, T.D., 1977, "Soil Movement Effect on Adjacent Public Facilities," ASCE Annual Convention, San Francisco, CA.

Otto, G.H., 1942. "An Interpretation of the Glacial Stratigraphy of the City of Chicago," Ph.D. dissertation, University of Chicago, Chicago, IL, 116 pp.

Otto, G.H., 1963, "Engineering Geology of the Chicago Area" Proceedings, Foundation Engineering in the Chicago Area, Soil Mechanics and Foundations Division, Illinois Section, ASCE, pp. 3-1 to 3-24.

Peck, R.B., 1969, "Deep Excavations and Tunneling in Soft Ground," State-of-the-art report, 7th International Conference of Soil Mechanics and Foundation Engineering, Mexico City, pp. 225-281.

Peck, R.B., and Berman, S., 1961, "Recent Practices for Foundations of High Buildings in Chicago," Symposium on the Design of High Buildings, Golden Jubilee Congress, University of Hong Kong, Hong Kong University Press, pp. 85-98.

APPENDIX II. CONVERSION TO SI UNITS

To convert	To	Multiply by
in.	mm	25.4
ft	m	0.305
lb/ft^2	kPa	0.0479

EXCAVATION AND SUPPORT SYSTEMS IN URBAN SETTINGS
J.P. Gould[1], G.J. Tamaro[1], J.P. Powers[2]

ABSTRACT

Factors that characterize the urban setting are: proximity of adjacent structures; intense prior use of the site; difficulty of access and limited work area; ubiquitous utilities and buried facilities; and special vulnerability to disturbance caused by the excavation and its dewatering. This paper summarizes in nine parts topics related to excavation and support systems for urban infrastructure. The first sections concern problems created by utilities and structures which interfere with the excavation. Then follows sections on dewatering in the urban setting, its cost and undesirable side effects and on the disposal of contaminated soil and groundwater. The final portions concern several specific geologic conditions which affect excavations in North Atlantic coastal cities.

INTRODUCTION

Excavation support systems are more difficult to design and install in cities than are retaining systems removed from the network of utilities and structures associated with the urban environment. City streets are congested with utilities ranging from water, electric, gas and telephone through services such as fiberoptics and private alarm signal controls. Urban sites are also encumbered by overhead bridges, utility tunnels, pedestrian tunnels and adjacent structures, occasionally covered by elevated rail or highway structures and either intersected or paralleled by rail or highway tunnels. The designer must thus consider the difficulty of installation of the earth support system as well as the effect of installation procedures on existing facilities.

1-Partners and 2-Consultant, Mueser Rutledge Consulting Engineers, 708 Third Avenue, New York, New York, 10017

Excavation support must be installed within available rights-of-way and minimize lateral movements of the adjoining ground, utilities and structures. Substantial difficulties can be encountered when excavations are opened beneath existing underground structures such as rail or highway tunnels. Such excavations require both parallel wall systems installed under the intersecting tunnels as well as temporary or permanent underpinning of these structures. The choice of support is dictated by economics, safety and constructability. For a variety of reasons, the optimum system may not be selected by the contractor. The contractor may install a less conservative system requiring special care to satisfy the intended purpose.

MAINTAINING UTILITY SERVICES

Most urban sites are laced with a variety of public and private utilities, usually positioned at shallow depth and in reasonably predictable locations. Major collector sanitary and storm sewers can be much deeper, depending upon the geology, topography and flow conditions. Public facilities are in the street as a governmental right, while private utilities are in the street at the option of public agencies. The contractor must provide an earth support system that will either maintain the utilities' function or he must relocate the utilities to continue their service. Some utilities such as gas and low pressure water are operated in "loops" which can be temporarily interrupted if necessary. High pressure water lines usually are unique within an area and must be retained in position, are often vulnerable to ground movement, and, if of great age, highly sensitive to vibration or surcharge loading. Low pressure water or gas lines usually can be relocated and that cost is compensated by savings realized by not having to support the utility adjacent to or across the excavation.

Electric lines can be of low or high voltage. The former usually are easily handled either parallel to or crossing the excavation. In either case, the duct banks can be broken out, the cables bundled and temporarily supported (Figure 1). High voltage systems, such as "oil-o-static" lines, are difficult to protect inasmuch as the oil-containing pipe can be damaged by movement. Oil-o-static lines cannot be easily moved from side to side to facilitate installation of the earth support system nor can they be collected into a utility corridor and supported across the excavation. Telephone and fiberoptic lines are highly sensitive to movement, can be easily damaged by construction equipment and are difficult to repair. Large duct banks containing telephone lines

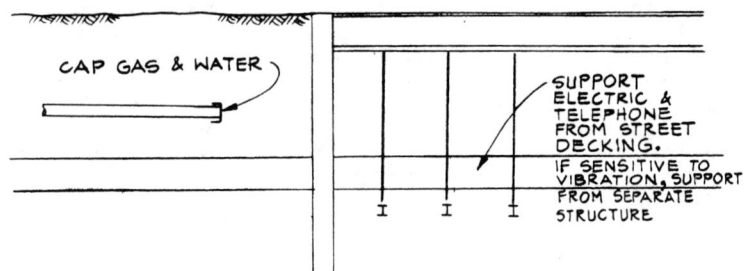

FIG. 1 UTILITIES TRANSVERSE TO WALL

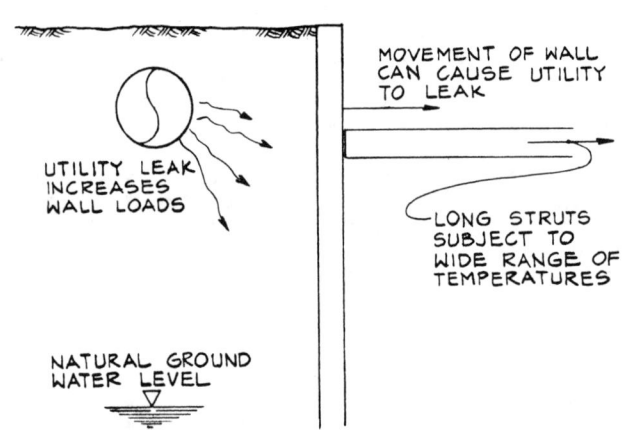

FIG. 2 DAMAGE TO PARALLEL UTILITIES

cannot be easily relocated since the numerous splices and reconnections of telephone or fiberoptic lines are time consuming and can potentially delay work, occasionally in the order of a year or more. It is best to work around these utilities in place and protect them with great care.

Sanitary, storm and combined sewers can be a serious problem to excavation support. Often they are of great age, sometimes constructed of brick or unreinforced concrete, are vulnerable to ground movements and can leak with the slightest provocation to saturate the surrounding ground (Figure 2). Leaking sewers have caused catastrophic failures either through overload of the wall or its supports or by loss of ground and resulting ground movements outside the wall leading to collapse of the earth retention (Figure 3). Sewers require great care in investigation and protection during construction. The outcome of minor leaks can be catastrophic. Utilities such as steam pipes, pneumatic lines, fire alarms, traffic signal controls are less extensive. High pressure steam pipes can be particularly sensitive to ground movement and should be treated with great care.

Regardless of how the main utilities are handled for the excavation, it usually is necessary to maintain basic services to adjacent structures at all times. Temporary outages, if accepted, are only permitted in off-peak periods to permit relocation or reconstruction of building services connections. It often is found expedient, where there is room between the existing structures and the future excavation, to relocate all of the parallel utilities into corridors inside the excavation. These new relocated utilities are usually less susceptible to damage from construction operations and movements and can provide continuing building service (Figure 4).

Investigations and Record Keeping

Before construction the designer and contractor should investigate available utility records and prepare composite drawings showing all information obtained from these records. The utilities should be identified on site to the extent of painting their positions on the pavement before construction. Test pits should be dug to verify that critical utilities are in the location indicated. Similar procedures should be followed for affected buildings. All existing records of overhead, below grade and adjacent structures should be investigated to determine the location and nature of foundations and the sensitivity of these structures to ground movements.

FIG. 3 COLLAPSE OF LARGE DIAMETER SEWER DUE TO SEEPAGE & GROUND MOVEMENT

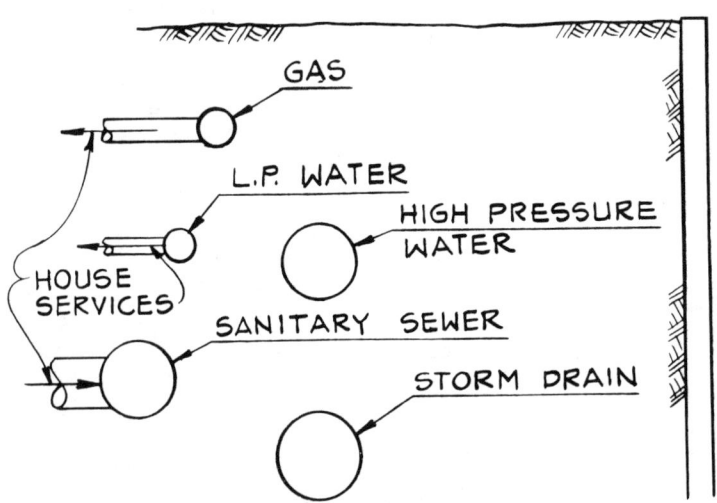

FIG. 4 RELOCATE UTILITIES PARALLEL TO WALL OUTSIDE LIMITS OF EXCAVATION

An existing condition survey and an optical survey of all utilities and buildings should be performed prior to construction. The contractor relocating utilities during construction should maintain an accurate record of the relocated position. Background levels of noise and vibration should also be determined before the start of work. The monitoring program should continue for a sufficient period after construction to assure that the utilities and structures have stabilized and that no further movements are occurring. At that time a final condition survey is performed to establish that damages have not occurred to the structures and to protect against damage claims.

PROTECTING ADJACENT STRUCTURES

Excavations at urban sites frequently must be undertaken adjacent to or beneath buildings and elevated transportation structures such as pedestrian bridges, road or rail ramps. Foundations for these structures usually lie within the bed of the street and require temporary support during construction. At some locations right-of-way restrictions may require installation of an earth retention system under such a structure. Underpinning methods or sloping wall systems may be satisfactory in those cases (Figure 5). Below the street are pedestrian, highway, rail, or utility tunnels. Utility or pedestrian tunnels are usually transverse to the street line while transit tunnels usually parallel the alignment and create a continuing problem for the full length of excavation.

If the foundations for the overhead structures are shallow and fall outside the excavation they can be protected either by nominal underpinning or a rigid ground support system. These structures, however, may provide overhead clearance problems during installation of the excavation support. If the overhead structures' foundations lie within the excavation limits, it is necessary to provide temporary vertical support within the excavation, to carry the foundation to a level below subgrade of the excavation or to provide transfer structures to carry the load on the earth support system (Figure 6).

In some cities, local authorities require temporary roadways to provide passage of traffic through the site. In these cases it is necessary to install temporary foundations either in conjunction with the retention system or within the excavation to support these temporary overhead structures, (Figure 7). Structures perpendicular to the length of the excavation are usually handled by

FIG. 5 BORED PILE WALL INSTALLED AT ANGLE BELOW BUILDING

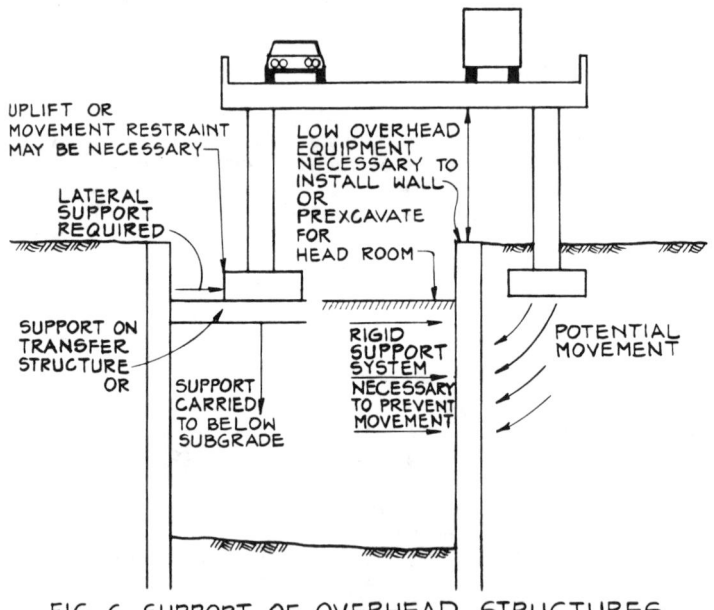

FIG. 6 SUPPORT OF OVERHEAD STRUCTURES

temporary supports, temporary underpinning, permanent underpinning or by relocation.

Large diameter sewers, or transit or highway tunnels located at depth beneath the excavation can also pose difficulties. In many cases these structures are disturbed by removal of overburden pressures. A tunnel underlying an excavation in overconsolidated clay, for example, will have a tendency to heave. Swelling and heaving of the soil is usually time-dependent and sensitive to the length of time the excavation is open as well as exposure to water. Structures underlying excavations may be subject to buoyant forces which had been resisted by the soil to be excavated. Dewatering or special support methods may be necessary to prevent heave.

In many cases, removal of overburden or lateral restraint on an unreinforced structure, such as a large diameter sanitary interceptor, will allow the structure to deform in a manner contrary to its initial design and current equilibrium condition. These structures may begin to leak because of the changes in ground pressures and could even collapse. Existing brick sanitary or storm sewers may also prove to be a problem if the overburden or lateral support is removed from the sewer. Some of these facilities operate under fluid surcharge or are subject to tidal effects and may carry internal pressures which can exceed resisting soil pressures and, as a result, could rupture (Figure 8).

Selection of Earth Support System

The retention system selected for an excavation must take into consideration the location and nature of all utilities and structures adjacent to or within the excavation limits. Some systems are adaptable and generally more economical, but have certain short-comings in performance. For example, soldier pile and lagging walls have great adaptability in positioning soldier piles to avoid utilities. Lagging can be cut and fitted to accommodate variations dictated by utilities or buried structures. However, the soldier pile and lagging wall is yielding to some extent and often not suitable for a high water table, running sands or squeezing clays. Ease of installation could be offset easily by damage which may occur during excavation. Other systems such as steel sheeting or concrete slurry walls are less adaptable in accommodating utilities. Gaps in the wall may be dictated by the position of utilities which cannot be moved. Certain techniques have been developed for placing slurry walls under and around utilities (Figure 9). These methods require great care in installation and are best

FIG. 7 TEMPORY VEHICULAR OVERPASS OVER SUBWAY CONSTRUCTION

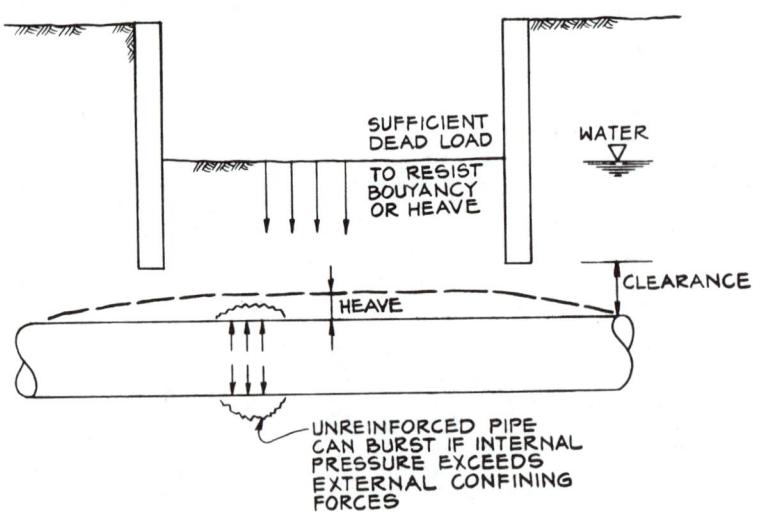

FIG. 8 UTILITIES OR STRUCTURES BENEATH EXCAVATION

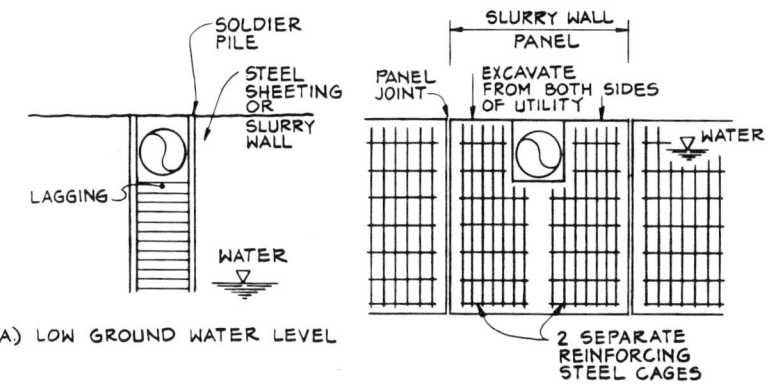

FIG. 9 WORKING AROUND UTILITIES

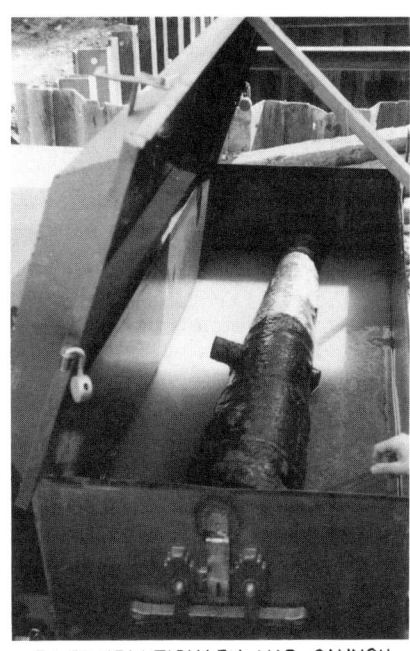

A.) ARCHEOLOGICAL "DIG"   B) REVOLUTIONARY WAR CANNON

FIG. 10 ARCHEOLOGICAL "FIND"

left to specialists. In some cases pit underpinning techniques must be used to install an earth retention system under an existing structure or utility. Noise and vibration considerations may also influence the selection of a system.

The choice of excavation support requires much forethought during design. The design engineer must consider all factors which will affect the performance of the earth support as well as how that temporary system may affect the permanent structure. Appropriate schemes should be studied and a preliminary design prepared to verify that the method to be recommended is constructible and will perform as intended. Clear direction should be given to the contractor to rule out inappropriate procedures and for protective measures which must be taken during excavation. The excavation support should then be designed and detailed by the contractor and closely inspected by the engineer during installation and excavation to assure that the contractor has met these standards.

ARCHEOLOGICAL AND ENVIRONMENTAL ISSUES

With increasing interest in the history of our cities, archeological investigations have become more prominent on urban construction projects. Often governmental agencies require an archeological evaluation of the site prior to start of work and stipulate on-site investigation. Without an archaeological study and a pre-construction archeological "dig", unexpected "finds" occur. Projects are delayed and the site is made unavailable for further work as a result of the discovery of "ancient" constructions, "ancient" burial grounds or a colonial privy. These archeological finds occur occasionally within streets during subway construction, such as the discovery of Henry Hudson's Tyjer hull in Manhattan in the early 1900's, or within sites, such as the recent discovery of a colonial burial ground within the site of a Federal Office Building in Manhattan or of four British cannon at another site (Figures 10a and 10b). Urban site archeology must be considered in advance since unanticipated finds can delay and impact the work. Interviews of local archeologists familiar with the history of the region or studies of local historical maps and documents can assist in understanding problems which may result from archeological finds. Finds in European cities have included the discovery of unexploded ordnance from recent conflicts.

Urban sites are increasingly found to contain contaminated groundwater and soils, petroleum products, hazardous wastes, sewage residues, methane from broken utilities or decomposition of organic materials and stray DC electric currents where it is used for traction power on commuter rail systems. Experience indicates that under certain conditions, pipe lines and other conductive structures have a tendency to deteriorate when exposed to stray DC currents. Many commuter rail systems use the ground and any conductor in their path as the return. Stray currents can rapidly erode unprotected steel structures or tieback anchors. Electrical isolation of these elements may be required.

Contaminated groundwater usually requires special treatment during exploration, excavation and disposal. Contamination may also occur as a result of leakage of fuel or chemicals used during construction. Specialized consulting engineering firms investigate sites for contaminated materials within the proposed excavation. Such investigations should be undertaken if there is any suggestion that prior uses could have contaminated the ground or if evidence developed during the geotechnical exploration that groundwater or soil samples were contaminated. Special excavation procedures then may be required. Disposal of contaminated soil and groundwater is discussed in subsequent sections.

Great care must be taken to protect against both explosion and asphyxiation from gas exposure during construction. Dangerous gas accumulation may result from broken utilities, from construction equipment and materials, or from the decomposition of organic materials in situ. Special ventilation or open deck systems may be required during cut-and-cover operations.

DISPOSAL OF EXCAVATED SPOIL

Spoil disposal in a crowded urban setting has always been a problem. However, costs have escalated in recent years with increased environmental concerns, complex regulations by authorities, and limited future use of disposal sites. Uncontaminated sands can of course be an asset, salable at a profit. If the spoil has been contaminated with petroleum products, volatile organics or other substances classified as hazardous, disposal in secure landfills can result in very high cost. The permitting process can be time consuming. The owner should retain appropriate control over contaminated disposal to avoid project delays and future liability.

Rubble, timber, refuse, metal, soft organics and other materials frequently encountered in urban excavations may not be classified as hazardous, but may be unsalable as fill. It may be necessary to transport them to a sanitary landfill. Spoil from slurry excavations is frequently over-wet and disposal of excess slurry at completion of the work can be troublesome.

DEWATERING FOR URBAN EXCAVATIONS: COST FACTORS

For excavations at urban sites space constraints usually demand vertical ground support. Soldier piles and timber lagging are often the most economical wall, considering direct cost. Material and labor costs are modest, and the method is convenient for working around utilities and obstructions. But soldier piles and lagging require dewatering, while diaphragm walls or sheet piling may not. In reckoning total cost, the expense of dewatering in an urban situation is affected by the following factors.

Soil Conditions

In Figure 11, an excavation partially penetrates a sand aquifer. With an ample thickness (x) of sand between subgrade and underlying clay, the water table usually can be lowered with relatively few high-yield wells, and excavation proceeds in the dry. In Figure 12, however, the excavation penetrates into the base clay. Dewatering for lagging support can be difficult. Many closely spaced wells are needed to lower the water table as close as practical to the clay. Even if wellpoints on five foot (1.5m) centers were used, some inflow will occur across the interface. Placement of lagging boards just above the interface requires care and skill.

Other considerations aside, conditions of Figure 11 favor lagging, while those of Figure 12 favor a diaphragm wall or other cutoff. If a diaphragm wall were used for Figure 11, interior dewatering would still be necessary unless the cutoff was keyed into clay, potentially costly depending on dimension (x). If a sand layer exists beneath the clay, Figure 12, pressure relief may be required to prevent heave, depending on dimensions (H) and (Y).

Groundwater Regime

When the excavation penetrates to clay as in Figure 12, pumped wells work best if recharge to the aquifer is remote. The schedule must provide enough pumping time to deplete storage, after which seepage at the interface can

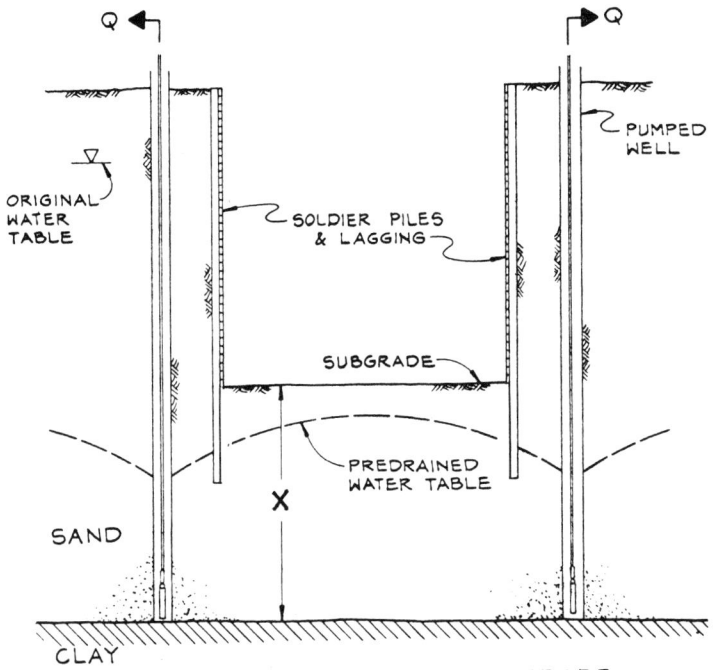

FIG. 11 SAND BELOW SUBGRADE

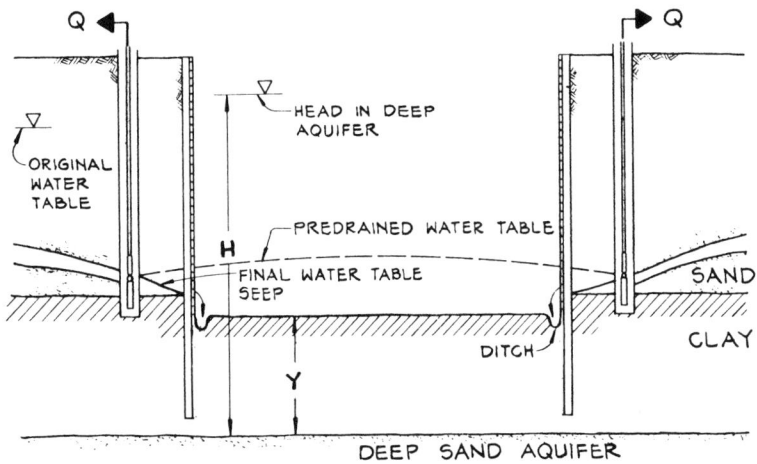

FIG. 12 EXCAVATION INTO CLAY

be reduced to a controllable amount. If there is nearby recharge, gradients between wells can be steep and more wells are necessary. Near sources in urban areas include:

- Surface water bodies connected to the excavation through sea walls, rock-filled cribs and porous bulkheads which had extended the shoreline.

- Leaking sewers or water mains.

- Gravel bedding around adjacent structures or utilities offers preferential seepage paths toward an excavation. Utilities laid in gravel can transmit water from great distances. If fines were piped from the natural soil during past construction, additional preferential paths aggravate conditions.

Disposal of Dewatering Discharge

In urban areas where combined sewers are connected to treatment plants, the receiving authority incurs a major cost in treating drainage water. Charges are now levied for accepting dewatering discharge, and projects which pump substantial flows for a significant period can face extraordinarily high disposal costs.

We are increasingly aware of contaminated groundwater and the impact of releasing it into the surface environment. The standards of regulating authorities must be met and obtaining discharge permits can be tedious and costly. Among the manmade contaminants which have had major impact on urban dewatering:

- Gasoline and other hydrocarbons that have leaked from service stations, fuel storage and refineries.

- Volatile organic solvents, leaching from waste lagoons at metalworking plants or released by dry cleaning establishments.

- Acid wastes, heavy metals and other materials spilled at chemical or industrial plants.

- In two large urban subways natural hydrogen sulfide in the groundwater has had major impact on groundwater control costs.

Methods for treatment of contaminated discharge include:

- For hydrocarbons and volatile organics: air stripping, carbon filtration or a combination of both. Disposal of spent carbon is costly.

- For acid waste: neutralization, sometimes followed by filtration, for which sludge disposal can be costly.

- For heavy metals or other undesirable solutes: precipitation with reagents, followed by filtration. Sludge disposal can be critical.

- For hydrogen sulfide: for one subway project the gas with its noxious odor was oxidized at moderate cost to water and elemental sulfur using hydrogen peroxide. On another metro project the regulators insisted the sulfide be oxidized to sulfate. The peroxide consumption was higher by a factor of five, other reagents were required, and the necessary treatment plant had a first cost of over a million dollars.

Union Work Rules

Modern dewatering systems, for the most part, operate automatically. Pumped well systems use submersible pumps requiring little maintenance. The old self-priming contractor's pump has been replaced with the automatic electric submersible. Wellpoint systems with high capacity vacuum pumps do not require tuning nor continuous attendance during operation. Nevertheless, in some labor jurisdictions archaic work rules are still enforced. The cost of operators around the clock is very high and can dictate the choice of dewatering method or use of an initially more expensive cut-off wall.

SIDE EFFECTS OF DEWATERING IN THE URBAN SETTING

Under certain conditions construction dewatering can have undesirable side effects leading to third party claims. Among these difficulties are damage to adjacent structures or to the new structure, and harmful impact on groundwater or surface water regimes. In the past we have seen misunderstanding on the potential effects, and instances of extraordinary protective measures that were probably unnecessary. In the last ten years advances in groundwater analysis techniques, and in understanding of geotechnical mechanisms involved in producing the side effects, have made the phenomena more predictable. The Underground Technology Research Council publication, "Dewatering - Avoiding Its Unwanted Side Effects" (1), addresses methods of analysis and mitigation. A brief description of the problems follows.

Damage Due to Improper Dewatering

Modern technology has provided new methods of groundwater control and has improved traditional ones. Still we have instances where an inappropriate method is chosen, or where a method that could be effective is applied harmfully. Powers (2) describes factors to consider when choosing and applying a method.

In Figure 12, a predrainage system of pumped wells or wellpoints can lower the water close enough to the clay so that remaining inflow can be handled by sumps, and the lowest boards installed. But if the predrainage is insufficient, or the lagging crews are inexperienced, ground can be lost. The difficulty is aggravated if the sand above the clay is clean and single-sized. Piping of fines through openings or into improperly filtered wells or sumps can cause gradual and unnoticed loss of ground leading to heavy damage to adjacent facilities.

In Fig 11, an effective predrainage system can lower the water table below subgrade. But if the system fails to decrease the upward gradient of pore water pressures to less than one, boils in the bottom can occur, harming the bearing properties of the foundation soils. Side effects due to improper dewatering can be avoided by appropriate specifications, effectively enforced, as described in Powers (2).

Settlement Due to Increase in Effective Stress

Lowering the water table shifts pore pressures to effective stress in the subsoils. The stress increment is modest, 62 psf per foot of drawdown (1 ton per square meter per meter). Older strata at depth generally can absorb the load without harmful consolidation, but recent compressible clays or organics nearby can experience significant settlement. The magnitude of consolidation depends on the prestress and permeability of the deposit, the magnitude and duration of drawdown. Resulting damage depends on the types of nearby structures and their foundations. Evaluation of the potential for damage must consider the previous history of groundwater fluctuations in the area.

Concern over damage from increase in effective stress usually tends to be exaggerated. Where an investigation indicates significant risk, the groundwater control specification can require excavation within cutoffs, limiting exterior drawdown. Alternatively, drawdown can be restricted by partial penetration of the dewatering wells, by artificial recharge or by creating a barrier of decreased permeability. On some projects the most cost-effective solution has been to accept limited damage, and pay for its repair.

Buried Timber Structures

Where untreated timber piles or other buried wood structures are present below the water table, they are to some degree naturally protected. If the water table is lowered the timber may be exposed to aerobic organisms which can cause deterioration, sometimes quite rapid. Where the piles are imbedded in silts or clays, the problem is less severe, since fine grained soils do not lose much moisture or gain much oxygen during temporary dewatering. If, however, the piles are imbedded in free draining soils and fill oxygen can reach the surface of the timber when the water table is lowered. Then the aerobic organisms can thrive and deterioration may occur. It may be advisable to severely restrict dewatering, or to inject water near the piles to prevent their aeration.

Adjacent Groundwater Supplies

Where the aquifer to be dewatered is being exploited for water supply, the supply wells may experience temporary loss of capacity. The aquifer can suffer long term damage if it is overpumped for dewatering. In coastal areas salt water intrusion is a risk. In urban areas where plumes of contaminated groundwater exist, their movement can be accelerated by dewatering. Problems have been mitigated by providing a temporary or permanent alternative water supply, by excavation within cutoffs, by restricting wells to partial penetration, and by artificial recharge.

Trees in Urban Parks

Concern has been expressed over the effect of long term dewatering on vegetation. Classic examples involved in urban excavations are the 300-year old trees in Harvard Yard and the trees along the Capitol Mall in Washington, DC. Unless the trees are hydrophilic, with roots reaching the water table, the concern may be unwarranted. Excessive irrigation can do more harm than good. A recommended procedure is to retain the services of a qualified botanist to monitor the condition of the trees during dewatering. The botanist will be expected to direct corrective action to maintain the desirable moisture balance.

Wetlands

Wetlands exist near many of our urban areas, and can be a subject of concern. A wetland may be described as a land area partly covered with shallow water, or subject to intermittent flooding and slow drainage. A more precise definition is controversial, and currently the subject of dispute among environmental groups and agencies of the Federal government.

Good et al (6) describe the complex ecosystems in various types of wetlands, and the surface and groundwater hydrology that affect them. In its different zones and at different seasons typical wetland may experience groundwater recharge or groundwater discharge. A dewatering system operating near a wetland may temporarily alter natural conditions. When the wetland is in its discharge cycle, flow from springs may diminish. Whether the impact of construction dewatering is greater than normal cyclic fluctuation, to which the ecosystem is adapted, must be evaluated. Release of fresh groundwater discharge from a dewatering system into a salt water marsh, or brackish water into a fresh water marsh, might also upset normal conditions.

Before dewatering near wetlands, it is recommended that a qualified specialist be consulted. It may be concluded that the dewatering operation will not have significant effect, or that special measures should be considered. A specialist can determine measures to give reasonable protection to the wetland without unduly increasing project costs.

EFFECT OF GEOLOGIC UNCONFORMITIES

To focus attention on excavation problems characteristic of cities of the North Atlantic coast, the effects of certain common geologic features is discussed in this section. There are two major unconformities in their subsurface profiles: Pleistocene sediments over bedrock in Boston, New York, Philadelphia, Baltimore and Washington; and Pleistocene over Cretaceous Coastal Plain sediments in Washington and Baltimore and in New York City to a lesser extent. The first of these discontinuities is characterized by:

- Relatively coarse stream terrace or glacial outwash deposits overlying bedrock.

- High permeability at the base of the overburden with a sudden decrease in permeability as the profile passes into rock.

- Weathering, softening and jointing of the upper portion of bedrock.
- An abrupt increase in strength and resistance to excavation passing across the interface.
- A special difficulty of defining material properties at either side of the interface: coarse, poorly-bound, overburden above, irregularly weathered bedrock below.

The unconformity of Pleistocene over Cretaceous is characterized by a sudden permeability decrease, strong anisotropy and highly directional permeability in the glacial deposits, and, at some locations, plowing and overturning of the ordinary sequence of strata near that boundary.

Surprises in excavating through recent overburden soil into rock are manifold. Where bedrock is deeply weathered as in Washington and Baltimore, it is common for test borings with powerful rotary drill or hollow stem-auger to be advanced completely through the decomposed and highly weathered bedrock, taking spoon samples with high blow counts all the way down. Then, contrary to the boring logs the material is discovered to act rock-like rather than soil-like in the actual excavation. Acrimony may follow, leading to a changed conditions claim.

Where potentially runny glacial sand or silt overlies an ice-scoured rock surface, as in Manhattan, irregularities in the rock are difficult to seal perfectly. Sheetpiling for the support wall is driven to the top of rock, leaving an imperfect fit between their tips and the rock surface. Exterior head can cause piping and loss of ground at low points beneath cofferdam walls which bear on rock.

Street Excavations Across the Bedrock Boundary

In a crowded urban setting, problems arise if the driven cofferdam wall meets refusal on the bedrock surface while the interior excavation continues below the tips of the sheeting or soldier piles. The tips of soldier piles remain supported on a narrow ledge of jointed bedrock. Stability is threatened by the vertical load of street decking or the vertical component of tiebacks on the soldier piles.

An example of the consequences was the failure of Section A003 cofferdam of the Washington Metro Subway on Connecticut Avenue in front of the Mayflower Hotel. A typical cross section is shown in Figure 13 with a photo of the collapsed street in Figure 14. This multiple-box structure was to be placed in cut-and-cover directly beneath the active thoroughfare with soldier piles supporting street decking. The deck served as a brace and with a single tieback resisted the lateral earth loading. The strike of metamorphic bedrock paralleled Connecticut Avenue, joints dipping between 45° and 60° west, with this unfavorable orientation exiting the east face of the cut. In 1972, when excavation was essentially complete a length of the east soldier pile wall failed by pushing out the rock wedges on which they were perched, causing the street decking to collapse with major disruption to civic activity and to the project.

Providing Stability

Cofferdam stability threatened in this manner can be maintained by: anchoring the rock wedge to provide horizontal restraint; drilling a heavy, near-vertical dowel through rock at the base of the soldier pile for direct vertical support; or by underpinning the wall elements. To illustrate the scale of the problem, Figure 15 shows the potential failure body created by this soldier pile on a rock wedge bounded by dipping joints. The graph summarizes the necessary horizontal stabilizing force, plotted against the dip angle of the bounding rock joints and the effective friction angle of those joints which reflects their roughness and shear keying. The wedge is assumed to be bounded by symmetrical joints centered on the back of the soldier pile. The weight of the wedge generally will be a small percentage of the pile load and is neglected. As the joint planes steepen, the necessary stabilizing force increases. The numerical effect of a change in dip angle nearly equals the influence of a change in friction angle. To just balance stability, a horizontal force of 1/4 to 1/3 of the pile vertical load would be needed at an unfavorable joint dip of 65° with fairly smooth, weathered joints having ϕ between 35° and 45°.

To foresee the Section A003 failure one would have to recognize the unfavorable attitude of the rock joints and their weathered condition directly below the rock surface. The geotechnical investigation should attempt to delineate the basic rock joint attitude, postulate a reasonably conservative friction angle and inform the contractor of the hazards that impend for soldier piles which refuse above subgrade.

URBAN EXCAVATION AND SUPPORT

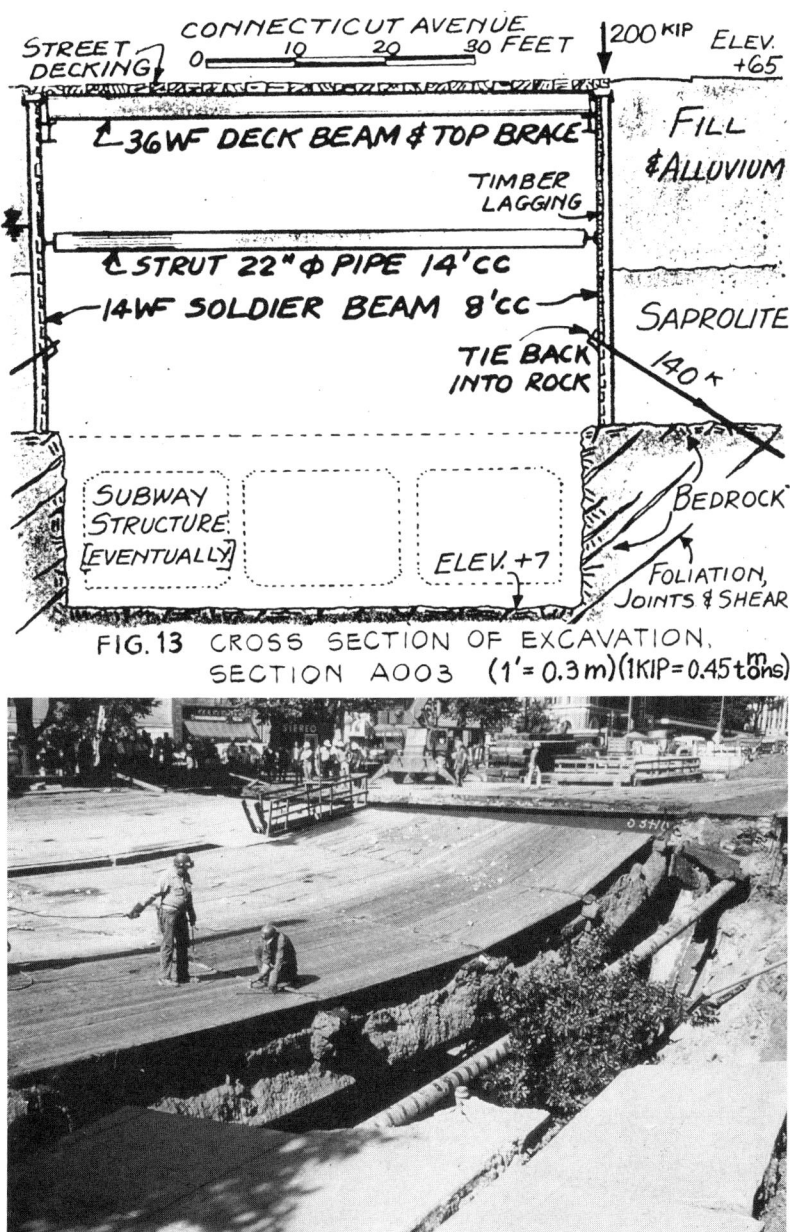

FIG. 13 CROSS SECTION OF EXCAVATION, SECTION A003 (1' = 0.3 m)(1KIP = 0.45 tons)

FIG. 14 FAILURE OF EXCAVATION, SECTION A003

Interface of Pleistocene over Cretaceous

New York City geology includes an important discontinuity between Pleistocene overburden and Cretaceous Coastal Plain sediments on Long Island and along the southern shore of Staten Island. At the latter location coarse grained glacial outwash or till overlies Cretaceous Magothy sands. The outwash is confined by overlying impervious till, and derives copious recharge from upland sources. The outwash is oriented in trains or troughs parallel to the southeast direction of glacial advance, presumably also the direction of the original draining melt-water. To add an unpleasant feature, there are locations where the plowing action of overriding ice has jumbled the outwash with the Cretaceous, reversing their order and implanting blocks of Cretaceous within and above outwash.

These conditions created an unexpected array of difficulties for tunnels and shafts of a major collector sewer crossing southern Staten Island along the geologic interface, as discussed by Wagner and White (3). After extensive field and laboratory investigations, the properties of the outwash sands and the Cretaceous were compared as follows:

	D50 Range Millimeters	Percent Minus #200 Sieve	Horizontal Coeff. of Permeability cm/sec
Outwash Sand and Gravel	0.5 to 1.1	4 to 10	1.5×10^{-1}
Cretaceous Magothy Sand	0.4 to 0.6	2 to 8	3×10^{-2}

The important characteristic is the five to one decrease of permeability across the interface. The greater horizontal permeability of the outwash results from its strong anisotropy and the sorting of gravelly streaks lying in dips on the interface. Shafts and tunnels passed through this level where water flowed in along the top of the Cretaceous without being able to drain down into the less pervious sands underlying. At individual shafts, up to 9000 gpm (34 cubic meters per minute) were pumped in order to excavate and install circular liner plate and ring beams. Often exceptional pumping did not suffice and a variety of auxiliary measures were necessary: installing internal sheetpiling or jet grouted walls across the interface, extensive intrusion grouting or dewatering in wellpoints lancing outward along the interface. Even with these costly efforts, there was extensive ground loss, undermining the jacked sewer pipe in one location, necessitating air pressure in another tunnel section and extensive revisions in both shaft sinking and tunneling procedures.

Interestingly, without direct glacial intervention, similar difficulties have been encountered at this same interface of Pleistocene terrace deposits over Cretaceous strata in Washington, DC. Here, the Coastal Plain sediments are extensive and variegated and the unconformity contacts a range of soil types, usually with the overlying stream terrace materials several times more permeable than the Cretaceous below. This interface often forms an undulating surface upon which seepage into excavations is concentrated at low points on the top of the Cretaceous with the attendant difficulties of soil piping and loss of ground.

SPECIAL PROBLEM SOILS

A range of soils, soft organic silt, bouldery outwash and slickensided clays pose problems to installation of retaining structures. Possibly the most treacherous is the classic clean, single-size sand and non-plastic rock-flour silt. These have the characteristic mobility which makes them unstable from vibrations or from seepage forces directed toward the excavation. The most troublesome range of sands are those clean, uniform materials noted for their liquefaction potential, Figure 16. Also included are glacial lake varved silts, the notorious "bulls liver" of Metropolitan New York. The glacial lake sediments are overlain by delta sands marking the final stage of deposition. These deposits are concentrated in Manhattan on the Lower East Side, in a central zone between Foley Square and 30th Street and in the Harlem lowland. Although preloaded by ice and of low compressibility under static load, they have a bad reputation for instability under seepage forces. The delta sands densify significantly under construction vibrations with consequent settlement. Other American cities have their own examples of potentially unstable single-sized Pleistocene sands. For example: well-sorted lenses in the Los Angeles River alluvium; and in stream terraces built up by Potomac flood waters in Pleistocene times, in Washington, D.C.

Difficulties in the Excavation

An example of the problems created by the runny character of such sands is the settlement of surrounding facilities caused during installation of anchor ties for

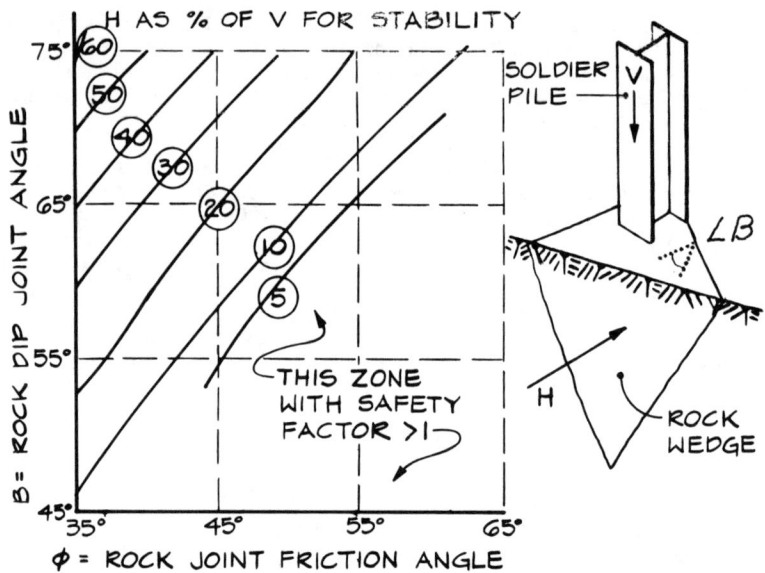

FIGURE 15, SOLDIER PILE ON ROCK LEDGE

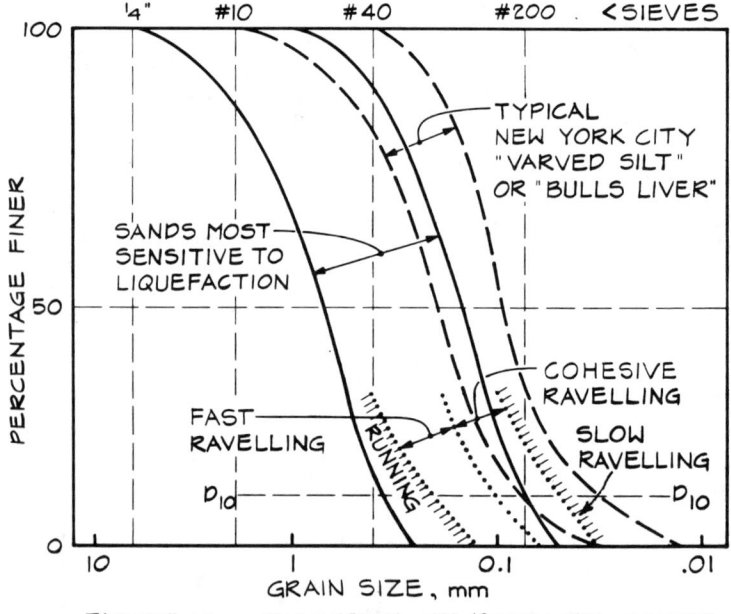

FIGURE 16, GRADATION OF PROBLEM SANDS

the slurry wall of Manhattan's World Trade Center in 1969. The piping of glacial fine sand and silt through the anchor holes created such severe loss of ground that a truck broke through the street pavement at one location and the adjacent subway required extensive track re-leveling. This is a typical problem with tiebacks installed in cohesionless sand where a drill casing is advanced by washing ahead or by washing at the end of the casing with water returning up outside the casing.

In lower Manhattan there has been a series of troublesome episodes caused by vibrations from driving bearing piles, soldier piles or sheeting through those same delta sands. Their gradations lie within the range of liquefiable sands, but being ice-loaded they are medium compact with SPT N values in the range of 15 to 30. The vibration level is low, typically less than 0.5 feet per second (0.15 mps) particle velocity. But long-continued driving of many piles for a large building foundation superposes the very small shakedown of a very large number of driving episodes. The cumulative effect has been several inches of settlement of peripheral areas, occurring on a steep gradient away from the excavation, as discussed by Lacy and Gould (4).

In Los Angeles, the new subway curving from Hill to 7th Street tunneled beneath two blocks of buildings on spread foundations in alluvium above the water table. Compaction grouting from a large number of pre-installed borings was intended to compensate for tunneling ground loss in a zone 10 to 20 feet (3 to 6m) thick of alluvium between the tunnels and the building footings. The mass of the alluvium was so compact as to prevent expansion of the grout bulb. The real cause of difficulty proved to be the presence within the alluvium of clean, uniform, runny sand in scattered lenses which threatened to cave without warning from the crown and top heading. Compaction grouting could not be invoked compensate for an unpredictable and erratic loss of ground from beneath the footings. The contractor's solution was to set up a plant within the tunnel for sodium silicate pattern grouting of the crown ahead of the face. Combined with careful mining, this permitted the tunnel to pass beneath the overlying structures with only a few millimeters of settlement of their footings, as described by Robison and Wardwell (5).

The first U.S. experience with the NATM tunneling in soft ground was at the Washington Metro Subway Fort Totten Station. The 60 ft. station width was opened with five successive drifts, carefully closing the face and crown with shotcrete after each advance between installation of lattice girders. As in the Los Angeles case, difficulties arose from the presence of clean runny sand pockets within the dense and generally stable Cretaceous formation. The contractor protected the heading by grouting as the need was indicated from within the tunnels.

Controversy arose when the Metro Section D004 tunnels passed beneath the 10th Street Mall bridge within Pleistocene terrace sands and caused 4 to 6 inches of settlement due to caving from a poorly breasted face. In this case, the Contractor's consultant recovered sand samples by horizontal scraping from the tunnel face. Gradations of these samples were compared to pre-bid gradation tests on samples recovered in conventional split spoons, crossing the horizontal lenses and laminations in the vertical direction. These spoon samples mixed the successive layers and averaged the silty and clean sands. The contractor was able to show that this discrepancy placed the in-situ sands in the range of running ground as compared to the stable silty sands implied in the pre-bid data and, thus, successfully pressed his claim of changed conditions. The consultant's distinction of the gradation of running and ravelling sands is included in Figure 16, and forms a spectrum passing through the range of sands subject to liquefaction.

SUMMARY

This paper reviews a variety of problems relating to urban excavations, the difficulties with utilities present on the site, the protection of these utilities and adjacent structures, problems of dewatering and the experiences with difficult geological conditions in East Coast cities. The emphasis is on those special factors in the urban setting which complicate the ordinary physical hazards of excavation. The increase in environmental and liability considerations has thrust on the designer and constructor a standard of care which is particularly exacting in the urban setting.

REFERENCES

1. Powers, J. P., Editor, 1985, <u>Dewatering - Avoiding its Unwanted Side Effects</u>, American Society of Civil Engineers, New York, NY, 69 pages.

2. Powers, J. P., 1992, <u>Construction Dewatering - New Methods and Applications</u>, Second Edition, John Wiley & Sons, Inc. New York, NY, 492 pages.

3. Wagner, J. R., and White, C. P., 1991, "Soft Ground Tunneling on Staten Island", Proceedings, Rapid Excavation and Tunneling Conference, Society for Mining, Metallurgy, and Exploration, pages 549-559.

4. Lacy, H. S., and Gould, J. P., 1985, "Settlement from Pile Driving in Sands", Vibration Problems in Geotechnical Engineering, American Society for Civil Engineers, New York, NY, pages 152-173.

5. Robison, M. J., and Wardwell, S. R., 1991, "Chemical Grouting to Control Ground Losses and Settlements on Los Angeles Metro Rail Contract A146", Proceedings, Rapid Excavation and Tunneling Conference, Society for Mining, Metallurgy, and Exploration, pages 179-195.

6. Good, R.E., Whigham, D.F. and Simpson, R.L., 1978, Freshwater Wetlands, Academic Press, New York, NY.

TUNNELING IN THE URBAN ENVIRONMENT

Norman A. Nadel[1], Fellow, ASCE

Abstract

Cities, almost by definition, bring structures and people into close proximity. In an ideal world, one would prefer to build tunnels away from cities, but the tunnels must be built where they are needed, and this frequently means that they must be built in cities. The constraints imposed by the proximity of people and structures make the construction of tunnels in an urban environment significantly more difficult, more costly, and more time consuming than their counterparts which are constructed in less confining circumstance.

Introduction

Tunnels are constructed for many purposes, including the transportation of people and goods; the conveyance of potable water, sewage, and electricity; the conveyance of water for power generation; the storage of commodities; military security; and as in the case of the Superconducting Super Collider in Texas, to provide a path for subatomic particle beams. They vary materially in length, depth below the surface, and in cross section size. Perhaps most important, the geologic medium through which they are constructed also can vary from sound, hard, dry rock to unstable saturated soil.

It makes a material difference to the difficulty of construction, time of performance and the cost of a tunnel if, in addition to all of the other factors, it is to be constructed in an urban setting. Some of the obstacles to the construction process which result from an urban location are obvious. Others are more subtle.

In the following discussion, there is frequent reference to working conditions in New York City. This is

---

1 President, MacLean Grove & Company, Inc., Greenwich, CT

appropriate because New York offers an opportunity to examine virtually every impediment to the urban construction process that exists anywhere. In short, if it isn't a problem in New York, it isn't a problem anywhere!

Work Sites

The greatest part of the length of a tunnel exists only underground. Surface land is required ordinarily only at shaft sites. However, even small quantities of land typically are not readily available in cities. Usually, the land in cities is virtually all occupied by structures, roads and streets, or by public parks. Park land is sacrosanct and any attempt to use it, even temporarily, will generate substantial outcry and opposition. What land can be made available elsewhere is very expensive. The placing of shafts in city streets is far from ideal because of the impact on traffic, local residents, and the interference of subsurface utilities.

In cities, with these limitations driving work site selection, it is not surprising that shaft sites provided to tunnel contractors are frequently very small, even when they are located off city streets. Because there are usually few available sites, considerations such as optimizing the geological setting and mitigating the negative impact on neighboring businesses and residents necessarily must be low on the list of priorities.

The contractor's needs at the shaft work site are many. First, the shaft itself, and the area that is immediately adjacent to it, requires some of the available space. Then space is required for the hoisting equipment, muck handling facilities, compressor plant, waste water settling basins, offices for the engineer and the contractor, shops, change house facilities for the workers, equipment and material storage, and so forth. Most tunnelers would agree that a comfortably sized shaft site for a major tunnel job requires a minimum of three acres of ground. This much space is rarely available for an urban job. Expensive compromises are therefore required. For example, offsite storage and more frequent deliveries of materials may be required; office, shop, and warehouse space can be rented in nearby buildings; materials and equipment can be stored on parts of local streets; and so forth. But these measures all involve additional expense to the contractor. A job outside the city need not incur these costs.

Impact on Neighbors

A tunnel construction shaft impacts its neighbors in

a variety of ways. To some extent, the work always generates noise, dirt, and air pollution. Depending upon the details, there can also be ground vibration. Even minor vibration caused by blasting has been demonstrated to be a source of concern to neighbors, and it is possible that adjacent structures can be damaged by substantial vibration. Settlement that results from loss of ground or dewatering can also damage nearby structures. Construction operations bring with them an increase in truck traffic on adjacent streets and roads, and finally, one can not overlook the effect to the ambiance of a neighborhood when a construction operation moves in. Those involved with construction may think a construction site is beautiful, but few members of the general public agree!

Alongside a tunnel construction shaft in a city street was a large water main to which there was connected a fire hydrant. The existence of this facility constituted a major hazard. Should a break in the main occur, the shaft and tunnel below could be quickly flooded. Accordingly, at the beginning of the job, the contractor arranged to have the main bypassed and shut down. To accommodate work to be performed later on, the main and hydrant had to be removed, and the contractor proceeded to make the necessary arrangements which required approval by both the fire department and the water department. Then a contractor licensed to work on the water mains had to be hired. This was a process which took several weeks.

After the main had been shut off, but before the main and hydrant could be removed, the contractor started the shaft construction on a single shift basis. A truck crane used for the construction was routinely parked in front of the inoperable hydrant during and after work. Needless to say, construction of the shaft and tunnel in the city street in a residential area was not very popular with the neighbors, and the contractor was barraged with complaints, and negative newspaper articles and editorials. He persevered, continued his work excavating the shaft, and finally was able to complete his arrangements for removal of the water main and hydrant on a particular day.

The night before that day, responding to neighbors' complaints, a policeman ticketed the crane for parking next to the hydrant, and the next morning the contractor removed the hydrant! The newspapers had a field day! Who else, but a corrupt contractor could arrange to remove a hydrant because he got a ticket for parking in front of it?

In rural areas, it is unlikely that one will encounter underground structures or utility lines that will interfere with the work. In the city, it is unlikely that one won't have such encounters! This is particularly true of new subway structures. In the construction of the new express tracks for the Sixth Avenue Subway in New York, the new tunnel was immediately adjacent to six different existing and operating subway tracks. Among other problems this caused, it was necessary to coordinate blasting so that no trains were in the vicinity at the time of the blast! The contractor was not allowed to stop trains except in case of an emergency. Considering that headways on some of these tracks were as little as three minutes, this was no minor restriction. The difficulties were compounded by the fact that in some cases, it was actually necessary to remove part of the existing structures to build the new tunnel.

Tunnel construction differs from most other construction work because, for reasons of overall economy and the need to perform the work within a reasonable period of time, it is ordinarily conducted on a multi-shift basis. Out of the city, the neighbors are far away, and the negative impact of this work schedule is small or non-existent. On the other hand, nearby neighbors, especially if they are residential, object strenuously when work is done in the vicinity outside normal working hours. In the city, there are always nearby neighbors, and as a result, limitations upon the contractor's operations are usually imposed by law or by contract in order to minimize the enumerated nuisances. Furthermore, in the city, there are thousands of "resident engineers" directing the contractor's work! It's not easy to work for so many bosses! It has been found that the establishment of a program to keep the neighbors advised regarding construction operations pays dividends.

<u>Noise</u>

Construction operations are noisy. What a neighbor hears is dependent upon several factors, the principal ones being the nature of the construction operations and the distance of the neighbor from the source of the noise. In a city, of course, neighbors are not ordinarily very far way. Many, if not most, cities have ordinances that limit the level of noise from construction operations that is allowed at adjacent property lines. For example, in residential areas, the New York City Noise Control Code provides allowances as follows:

```
    Low Density Residential        7 AM to 10 PM - 60DB
                                  10 PM to  7 AM - 50DB

    High Density Residential       7 AM to 10 PM - 65DB
                                  10 PM to  7 AM - 55DB
```

Except by special dispensation, blasting is allowed only during daylight hours, and the noise level from blasting is limited to 95DB.

As a practical matter, some of these limitations are virtually unachievable. There have been cases where it is impossible to determine whether the construction noise levels are in compliance because the ambient noise from sources other than the construction are greater than the allowances. The noise from the construction is drowned out by the ambient noise and cannot be measured! Typically, the intensity of the attempted enforcement of these regulations is driven by the number and the origin of complaints, and the level of effort expended by the contractor to mitigate the emanation of noise from the operations. Under the best of circumstances, construction proceeds under a mutually uncomfortable accommodation between the contractor and the enforcer of the regulations. Apart from prohibiting actually measured noise that is above a certain level, regulations typically also prohibit or limit construction activities on nights and weekends irrespective of noise levels produced.

All too often, contractors are forced to submit bids without a clear idea of what the working hours will be, a contracting version of Russian Roulette. In recognition of the problem, however, there is a growing tendency on the part of owners to assume the risk by specifying optimistic allowable working hours with the understanding that the owner will pay for the increased costs of more restrictive conditions. In this way, contractors don't need to include contingencies in their bids and the owner derives the benefit if the contractor can work more hours.

Blasting

Blasting is a major source of conflict between tunnel constructors and local residents in a city. Realistically, the employment of modern blasting techniques and the ability to accurately measure vibration make it highly unlikely that damage will be caused. But this does little to assuage the fear and emotional response to blasting from neighbors. Somehow blasting is perceived as being a form of violence and inordinate danger.

The negative emotional response is not limited to humans. It is usually required that a whistle be blown before a blast, and again after the blast to signal the "all clear". We received a telephone call one day from an irate woman complaining that we failed to signal the "all clear" after a blast. If so, we broke the rule, but what harm had been done? Her dog had developed the habit of hiding under the bed when the blasting signal was sounded, and emerging after the blast and the "all clear" whistle. She couldn't get the dog to come out from under the bed because although there had been blast, there had been no "all clear" signal.

Another problem concerns the storage of explosives. A typical tunnel blast may require many hundreds of pounds of explosives, and there must be ample site storage to accommodate the blasting schedule. In New York City, regulations normally require that deliveries be made only in daylight hours. This has the effect of increasing the need for jobsite storage, especially in the case of a job operating more than one shift, but New York will not allow storage of more than 1000 pounds under any circumstances, unless the site is in the park or other area remote from the public. Moreover, magazines are required to be armored and must be located a considerable distance from other facilities, thereby increasing the need for site space. In the past, however, as in the case of noise regulators, when the ability to effectively build the tunnel is at stake, accommodations are reached, usually in the form of permitting nighttime deliveries.

Regulations in New York require that there be a powder watchman on duty at all times when there are explosives in the magazine. The powder watchman may perform no other duties. On one shift jobs, arrangements are usually made with the supplier to pick up any unused explosives at the end of the shift, thereby eliminating the need for two expensive union powder watchman on the night shifts. On the other hand, deliveries and pickups are not inexpensive. Special armored trucks manned by two teamsters are required, and trucks may not carry more than 1000 pounds. Moreover, detonators and powder cannot be carried in the same truck. By contrast, on a typical remote site, explosives in substantial quantities are simply stored in a locked building with the key under the control of the blaster. With all of the restrictions, it is not surprising that explosives cost many times as much in New York City as they do elsewhere.

Labor

Labor costs vary considerably from location to location depending upon many factors. Perhaps the most important is whether the labor is unionized or not. By and large, work in cities is more likely to be unionized than in remote locations, and pay rates are likely to be higher in cities. Union manning and work rules are also more likely to be more restrictive in cities. However, it is impossible to generalize, and each situation must be viewed on its own merits.

Typically, a large number of tunnel workers are migratory, and they relocate to follow the work. Many of them find cities to be unattractive places to live, and it is, accordingly, harder to get qualified people in cities than it is in remote locations. In the cities, workers must spend a substantial amount of time commuting to and from work. In the country, they are likely to live within a few minutes drive of the jobsite. In the cities, if workers drive their own cars, parking is difficult. In some cities, unions require that employers provide parking for their workers, not always a simple thing for an employer to do. This, of course, is not a difficulty in rural locations where land is more plentiful.

Crime and Security

As we all know, crime is rampant in our cities. A work site in a city is an attraction to criminals and a security program must be in place to protect the work, the property, the equipment, and the people at work. On the construction of the New York City Third Water Tunnel, the City provided pay items for security personnel. After several armed robberies and sniping incidents, the City gave the contractor a change order to hire armed guards!

On the Third Water Tunnel job, the contractor provided walky-talky radios to about twenty key supervisory people in order to facilitate the coordination of various operations. When the supervisors came off shift, they would place their radios in a common battery charging unit. One weekend, despite around-the-clock guard service, the contractor's office was burglarized, and the radios were stolen. However, one of the supervisors had failed to place his radio in the charger, and instead had put his radio in his desk drawer. The burglars missed it, and it was there on Monday morning. The contractor was able to listen to the thieves utilizing the stolen radios to coordinate their activities in the conduct of other burglaries.

Because of the density of the population surrounding a work site, it is also important that the work not be a danger to innocent neighbors. The lawyers will tell us that a construction work site can be an attractive nuisance, and we must assure that people, particularly children, are kept away, especially when work is suspended on nights and weekends. This is no simple task!

Methods

All of the above restrictions affect construction methods. In cities, contractors are likely to arrange for the performance of many semi-specialized activities by independent businesses. Thus, in cities, all but the most elemental equipment repair and maintenance functions tend to be done offsite by others. Qualified labor and working space at the jobsite are just too difficult to obtain, and they are very expensive. Where there are strict and expensive union work rules, the potential manning of a piece of equipment is an important consideration in the selection of equipment to do the job. For example, union rules in New York City require that two miners be assigned to every rock drill. A short, small size tunnel may ordinarily best be drilled using portable light weight jackleg drills. In New York, however, each jackleg will require two miners, and it may pay to employ an expensive self-contained drill jumbo. The drills are more powerful, fewer drills will be needed, and thus the manpower requirement will be reduced. A common method of building a drill jumbo entails mounting drills on an old dump truck. The truck is driven into the face for each round, usually not more than once a shift. But the truck requires a driver who does nothing else. Contractors have accordingly built jumbos that are not self propelled, and which must be moved into the face by another piece of equipment. But that piece of equipment is then available to perform other activities.

On most remote jobs, the concrete required is obtained from an on-site batching and mixing plant. In cities, ready mixed concrete is used almost exclusively. Land limitations make this essential. However, it is difficult and expensive to obtain ready mixed concrete on more than one shift. Ready mix plants typically operate one shift regardless of the desires of the tunnel contractor. In cases where continuous, around the clock, concrete lining may be possible, it may be decided, if ready mixed concrete is required, to use a different concreting method which will be performed on one shift only.

Costs

It is not surprising that it costs substantially more to build a tunnel in a city than in a remote location. However, the magnitude of the difference is not well understood. Some years ago, our firm had occasion to prepare a bid for an underground pumped storage power facility that was in a remote, non-union, location in the southeastern United States. After completing our analysis of job costs, out of curiosity, we imagined that the facility was located in New York City, and we made another estimate. We made no changes in methods, and the only difference was in labor rates and manning requirements. The average wage rate in New York was 2.5 times that in the remote location. The total labor cost was 6 times as much, which indicates that manning requirements more than doubled the work force! On a tunnel job we were doing in New York a few years ago, our project manager was asked by a visitor, "How many people work here?'. He responded, "About a third!"

Conclusions

Building a tunnel in a rural area is very different from building a tunnel in an urban location. While the techniques to be employed may be similar, the job is very different. Different skills are required. In the urban location, creating the underground opening is only part of the job. Dealing with the neighboring and confining environment of the city requires great effort, sensitivity, and skill. The tunnel builder who underestimates this difference does so at great peril.

FREQUENCY BASED CONTROL OF URBAN BLASTING

By Charles H. Dowding[1]

ABSTRACT

Control of blasting is dependent upon determination of dominant frequency by one of the three methods compared herein. Several case studies are presented to demonstrate the importance of this frequency calculation for close-in urban blasting. Comparisons and case studies are synthesized to present a suggested method for a frequency based vibration control, and to project future trends in instrumentation and project monitoring.

INTRODUCTION

This paper reviews the technical background and data for the development of a frequency based particle velocity limit to control blasting operations. Details are presented for three different methods to determine dominant vibration frequency: 1) visual determination of zero point crossings and calculation with 2) Fourier & 3) single degree of freedom (SDOF) response spectra. Special attention is devoted to consideration of close-in blasts which produce high particle velocities at high frequencies. Such frequency considerations are important for urban infrastructure construction because the typically high dominant frequency allows higher particle velocity.

---

[1] Member ASCE and Professor, Department of Civil Engineering, Northwestern University, Evanston, IL. USA 60208-3109 (708)491-4338

First the importance of frequency is demonstrated with single degree of freedom (SDOF) response spectra. Observed cases of threshold cracking are then related to the Office of Surface Mining's regulatory velocity-frequency bound. The mathematical relationship between Fourier and pseudo velocity SDOF response spectra is investigated to show their relative usefulness to determine dominant frequency. Calculation of dominant frequency and visual estimation are then compared for 17 urban construction blasts to determine the adequacy of visual estimation. Finally, practical considerations are highlighted for close-in blasting, such as prevention of gas pressure induced block movement.

SINGLE DEGREE OF FREEDOM (SDOF) RESPONSE SPECTRUM

It is well known that the dynamic response of structures can be modeled as single degree of freedom systems and that the response of a wide range of structures can be plotted as a pseudo velocity response spectrum (PVRS) (Hudson, 1979, Newmark and Hall, 1969, Veletsos and Newmark, 1964). Such an approach employs time histories of the ground motion to excite mathematical models of spring, mass, and dashpot systems, which are the analogues of structures with various natural frequencies. The computed pseudo velocity response of the structure is the relative displacement, δ, of the structure walls times the circular natural frequency of the structure, $2\pi f$, where f is a structure's natural frequency. Relative displacement is particularly important because it is directly proportional to wall strains, which produce cracking.

To show the importance of frequency in the determination of the potential for cracking, two spectra in Fig. 1 were calculated for comparison with those of case histories involving threshold cracking. Spectrum A was developed from the ground motions recorded 72 m (220 ft) away from a single 91-kg (200-lb) charge detonated in a typical bench blast hole in a limestone quarry. Spectrum B was developed from ground motions recorded 12 m (38 ft) away from a 0-to 9-ms delayed tunnel blast with a maximum charge in any one delay of 1.7 kg (3.8 lb). The quarry blast generated a peak radial particle velocity of 43 mm/s (1.7 in./sec)and the tunnel blast generated a peak radial

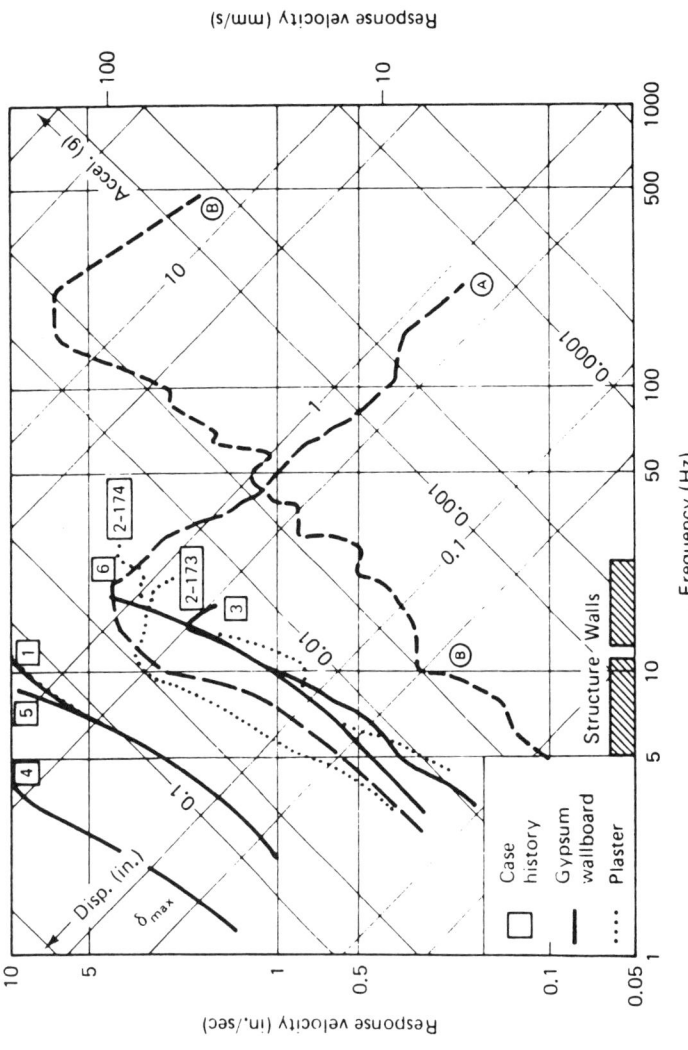

FIG. 1. Pseudo velocity response spectra of ground motions associated with cases of threshold cracking observed by the US Bureau of Mines compared with that for a tunnel construction blast (B) to show low relative displacements for high frequency motions in the range of structural response frequencies (5 to 20 Hz) (Dowding, 1985)

the solid line was taken from an appendix of U.S. Bureau of Mines (USBM) Report of Investigation, 8507, (Siskind et al, 1980). Unfortunately, the appendix was developed only as an initial proposal for such frequency control, not as a definitive closure (Siskind, 1987). Of even greater historical interest is the lack of data for a foundation for the German DIN 4150 standard. Numerous attempts by the Transportation and Road Research Laboratory in England to gain access to these data through official channels have been unsuccessful to date (New, 1991). Regardless of these drawbacks, some form of frequency control like that of the 8507 curve is necessary to control blasting operations in a manner that is equitable to both neighbors and excavation contractors.

It is instructive to explore the origin of the 8507 frequency bounds to access alternative means of determining dominant frequency. The general shape of the relationship was proposed by Swedish researchers associated with the International Standards Organization committee producing a blast vibration monitoring standard. Actual bounds between 5 and 30 Hz were then determined with cracking data measured by the USBM through calculation of Fourier frequency spectra from particle velocity time histories (Siskind,1987). These velocity time histories were obtained on the ground outside structures sustaining threshold cracks (Siskind et al, 1980).

The 8507 study concentrated on homes adjacent to mining facilities, and involved no direct measurement of large military or accidental blasts (dominant frequencies below 5 Hz) and very few construction blasts (dominant frequencies above 40 Hz). As shown in Fig. 3, the zones denoted by A and C are not well defined regarding the relationship between particle velocity and frequency and are in need of further research. As shown by the special control lines in Fig. 2, the author has employed several less restrictive bounds for close-in urban blasting. Within the frequency range of 4 to 30 Hz, the velocity and displacement bounds were defined with USBM data points shown in Fig. 3.

Table 1 summarizes information associated with the data points plotted in Fig. 3. Particle velocities are those in the horizontal direction as measured outside

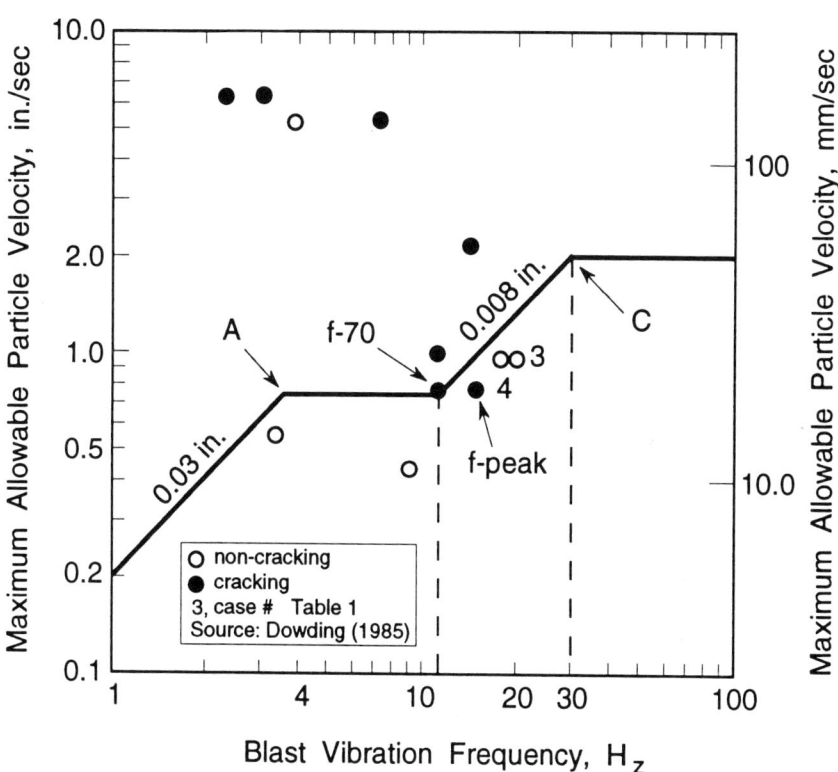

FIG. 3. Office of Surface Mining frequency-based velocity criteria compared with specific data points

TABLE 1
FREQUENCY CHARACTERISTICS OF CASE HISTORIES INVOLVING THRESHOLD CRACKING

Case #	f peak (Hz)	f−70 (Hz)	f+70 (Hz)	PPV (mm/s)	Shot #	Observation	CHD Case	Env
1	20	7	40	124 R	13	Crack	1	Test
2	30	13	40	24 T	173	Crack	2	Mining
*3	20	18	22	24 T	167	no crack	2	"
*4	14	12	17	20	10	crack	3	"
5	12	11	15	11	12	no crack	3	"
6	5	3	11	185	9	crack	4	"
7	7	5	8	136	8	no crack	4	"
8	35	8	40	302	2	crack	5	Construction
9	22	20	55	51	2	crack.	6	"
10	5	3.5	30	12	−	no crack	unpub	Mining

f peak = frequency at the maximum amplitude
f−70 = frequency at 70% of the maximum amplitude
* plotted in Figure 4, R = radial, T = transverse
CHD Case from Dowding, 1985

the house, and frequencies were calculated from SDOF pseudo velocity response spectra (PVRS). As explained later, PVRS can be compared directly with Fourier frequency spectra (FFS) when both employ velocity for input motions. PVRS for the two most critical cases in Fig. 3 (numbers 3 and 4) are compared in Fig. 4 to demonstrate how the dominant frequencies were determined. The frequency associated with the peak particle velocity is that of the peak of the response spectrum, f-peak, (Dowding, 1971) and the frequencies associated with f+70 and f-70 are those associated with response amplitudes equal to 70% of the peak. Plotted frequencies are f-70 as these are the most conservative approximation of the dominant frequency, lie on the displacement bound of the response spectrum and follow the method to establish the 8507 relationship.

Case numbers 3 and 4 from Table 1 are plotted as two points with both f-peak and f-70 in Fig. 3 to show the effect of two different strategies of frequency determination. Plotting with the lowest interpretable frequency, f-70, associated with the maximum particle velocity follows the procedure employed to establish the 8507 relationship. As shown in the figure, the 0.008 in. constant displacement bound is positioned between the two possible interpretations of cracking case 4 and produces the bound which is consistent with f-70 interpretation.

METHODS TO DETERMINE DOMINANT FREQUENCY

Despite the care taken by Siskind and his group to establish the bounds for the Appendix B curve in 8507, the scarcity of direct observations to substantiate corners A and C has contributed to ambivalence about the "BEST" method to determine dominant frequency. It was not felt that calculation of either Fourier or SDOF response spectra was necessary for all blasts when the vast majority of cases involve peak particle velocities below 0.8 ips (20 mm/s), the lowest particle velocity observed by the Bureau to cause threshold or hair sized cracks. Even for those blasts with larger peak particle velocities, many involve time histories where frequencies associated with the peak velocity are easily determined through visual inspection of the time history. As will be shown later, a controlled test of visual versus PVRS determination of dominant frequency showed that visual inspection was inappropriate for

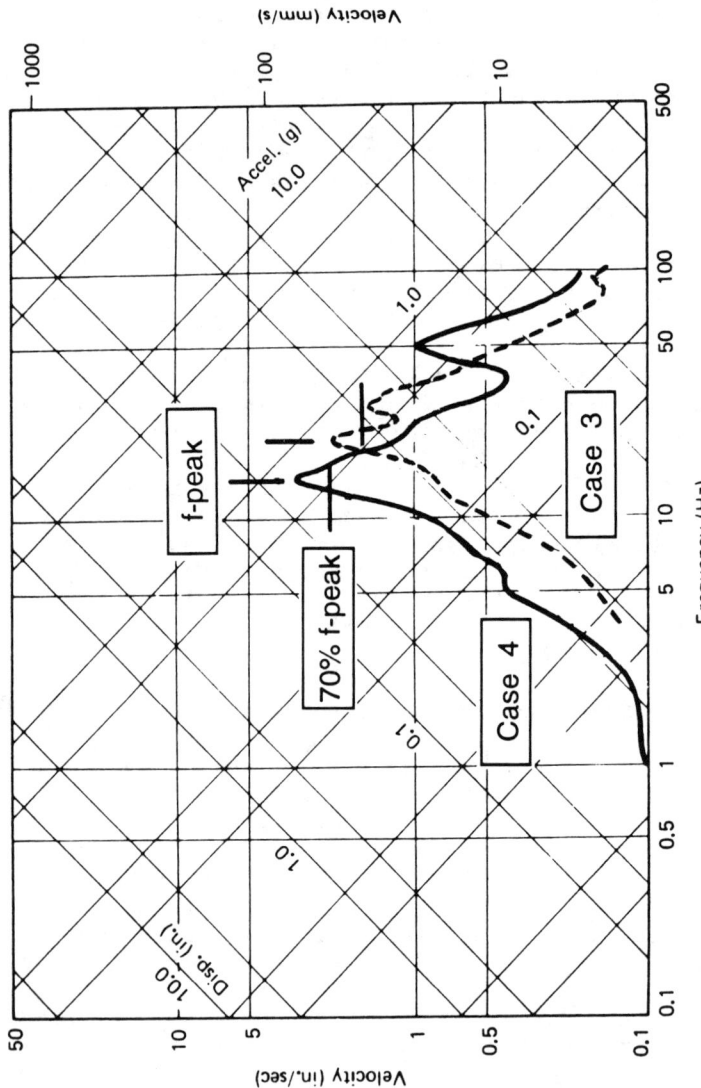

FIG. 4. Pseudo velocity response spectra of ground motions (points 3 & 4 in Figure 3) associated with threshold cracking that identify peak and 70% amplitudes and their respective amplitudes

only 2 out of 17 cases involving high frequency construction blasting, and in only one case was it unconservative. Thus for regulatory purposes, it is perhaps necessary to determine dominant frequency with a calculated spectrum for only 5 to 10% of monitored blasts.

As result of this observation that non-spectral determination might be sufficient for the majority of the cases, most instrument providers supply a frequency determination based upon a simple visual concept, the zero point crossing on either side of the peak. This calculation assumes the single peak is isolatable and is symmetrical as is the "peak amplitude B" in Fig. 5b. This frequency estimate can be supplied for all excursions in the time history or just for that of the peak particle velocity.

Thus there appears to be three principal means to estimate dominant frequency through calculation of, 1) Fourier frequency spectra, 2) SDOF response spectra, and 3) zero point crossing. The method of determination of PVRS spectra has been discussed already and that for calculation of Fourier spectra and zero point crossing will be discussed in subsequent sections. None are inherently superior, and it will be shown that the two spectra are essentially the same in their usefulness in calculation of dominant frequency. However, the SDOF pseudo velocity response spectrum is more useful for estimating structural response to transient excitation.

FOURIER FREQUENCY SPECTRA

Fourier frequency calculation is a popular method of automatically computing frequency content, not because of any inherent superiority, but because of its availability and historical use in solving mechanical vibration and signal processing problems (Bendat and Piersol, 1980).

It can be shown that amplitudes of the Fourier frequency spectrum (FFS) are equal to the SDOF <u>relative velocity</u> response spectrum with zero damping when the acceleration time history is employed as the ground motion input, Hudson (1979). Substitution of particle velocity for input in the Fourier equations reveals that the amplitude of the FFS is equivalent to that of

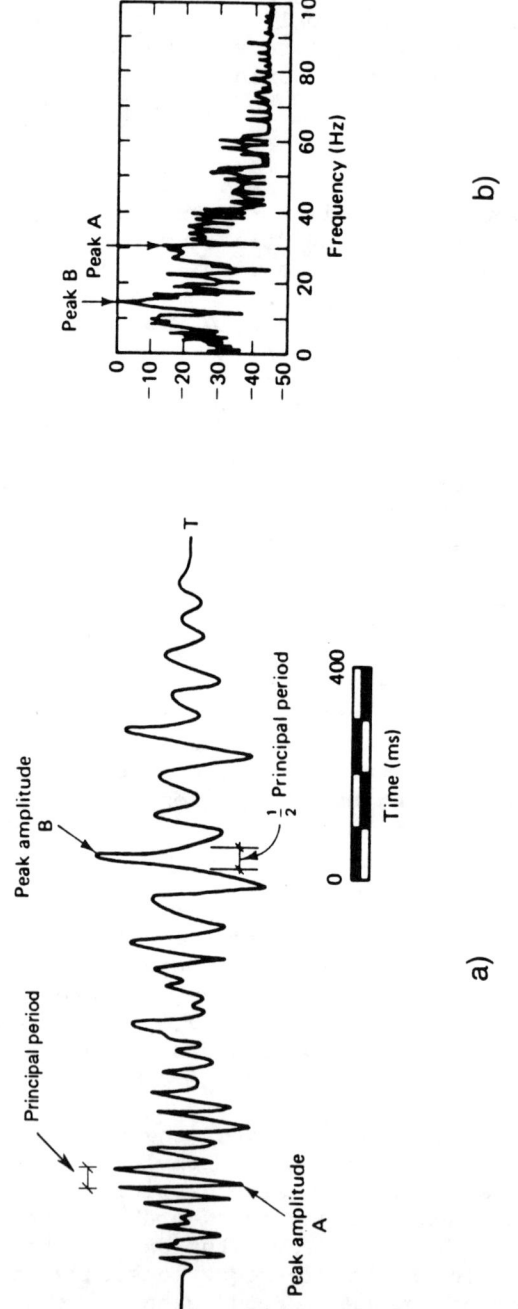

FIG. 5. Particle velocity time history from a surface coal mine blast showing the zero crossing points. The time history is shown as a) and the Fourier spectrum is shown as b)

the SDOF relative displacement response spectrum with zero damping. Thus, multiplication of the calculated FFS amplitudes by K $2\pi f$ will allow direct comparison with the pseudo velocity response spectrum (PVRS) with zero damping. The proportionality factor, K, is necessary to account for the reduction of relative displacement (and thus pseudo velocity) with damping for the response spectrum, and the $2\pi f$ is necessary to convert relative displacement to pseudo velocity. K is on the order of 1/3 for 5% damping.

Similarity of the FFS and PVRS determination of dominant frequency is illustrated by the comparison of calculated spectra of reciprocal machinery induced footing velocities shown in Fig. 6a. Motions are those in the single frequency dominated vertical direction and the less harmonic longitudinal direction. Fourier spectra (6b), plotted arithmetically, reveal the same distribution of dominant frequencies as do the pseudo velocity response spectra(6c), plotted logarithmically. Such similarity is expected as pseudo velocities are calculated by multiplying SDOF maximum relative displacement response by $2\pi f$ and Fourier amplitudes are already proportional to relative displacement. Spectral peak frequencies are the same for either FFS or PVRS. Frequencies associated with amplitudes 50% of the peak are 37 & 46 Hz for FFS and 34 & 51 Hz for PVRS for the vertical motions, and 16 & 48 and 17 and 55 Hz respectively for the longitudinal motions. Differences result from inclusion of damping in the calculation of the PVRS but not the FFS.

Comparisons such as those in Fig. 6 show that either Fourier or response spectra can be employed with equal mathematical validity for the automatic calculation of dominant frequency. However, standard Fourier techniques, developed for analysis of continuous phenomena, cannot be applied blindly to blast vibrations because of their transient nature (Dowding, 1988). Time histories must be machine digitized at frequencies at least 10 times that of the highest expected frequency of motion, not at the Nyquist frequency (equal to that of the expected) to capture peak motions. Proper units for the Fourier amplitudes are important and should relate to relative displacement, which is the source of structural strain and subsequent cracking. While pseudo velocity (for comparison with pseudo velocity SDOF response spectra)

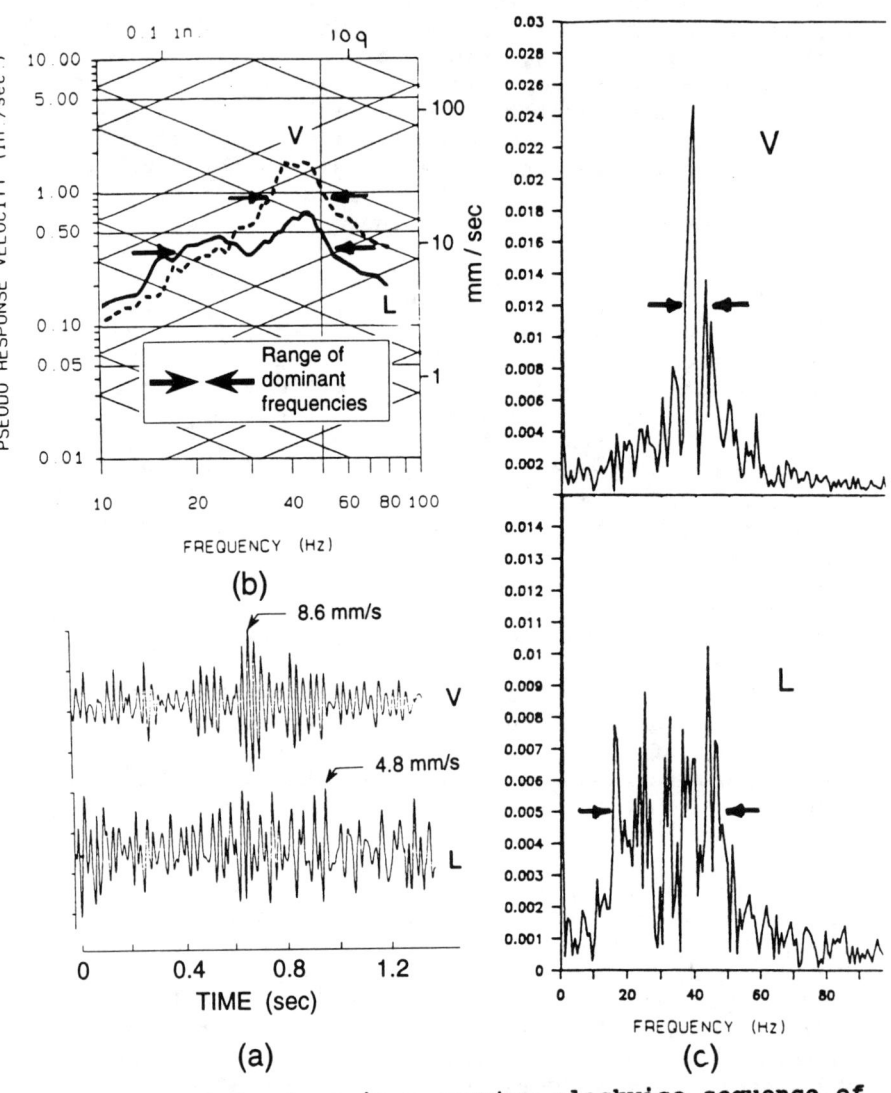

FIG. 6. Comparison, in a counter clockwise sequence, of particle velocity time histories (a), Fourier amplitude spectra (b), and single degree of freedom pseudo velocity response spectra (c) that shows the similarity of the two spectra

can be obtained by multiplying FFS amplitudes from velocity input by K $2\pi f$, Crenwelge (1987, 1991) has suggested that raw velocity be calculated from FFS by multiplying by $2(\pi)(\delta)f$, where $(\delta)f$ is the interval between frequencies for which the Fourier amplitude is calculated. In addition, Crenwelge (1991) has concluded that smoothing of the FFS amplitudes is necessary to eliminate spurious effects of "combined-amplitude-frequency component bandwidths". As he points out, this smoothing is acceptable because of the energy in the PRVS dissipation effects of structural damping that control structural response. Continued development of the FFS approach will no doubt bring about further modification and a growing similarity, if not equality, with the SDOF pseudo velocity response spectrum.

VISUAL AND/OR ZERO POINT CROSSING FREQUENCY DETERMINATION

There are several variations to the zero point crossing approach. In Fig. 5, peak amplitude B, the two positions where the time history crosses the zero amplitude line are shown and define the length of 1/2 of the principal period, T (= 1/f), which is shown in the accompanying Fourier frequency analysis to occur at a frequency of 17 Hz. At the higher (30 Hz) frequency peak, A, the distance between adjacent peaks defines the principal period. This peak to peak variation is most often employed for visual frequency estimation of high frequency waves with very small peak to peak distances.

Digital blast vibration monitoring equipment processes data in a form which allows automatic calculation of the frequency associated with the peak particle velocity amplitude. The method of this calculation should always be checked with the instrument manufacturer to ensure proper interpretation. If such automatic calculation can be made for the peak velocity excursion, it can be calculated for all excursions in a time history. Results of an excursion analysis for an entire time history are presented in Fig. 7 along with the associated time history. Such graphical presentations allow a visual comparison with the 8507 curve and ensure that a combination of a non peak velocity at an unusually low frequency does not exceed the frequency based criterion.

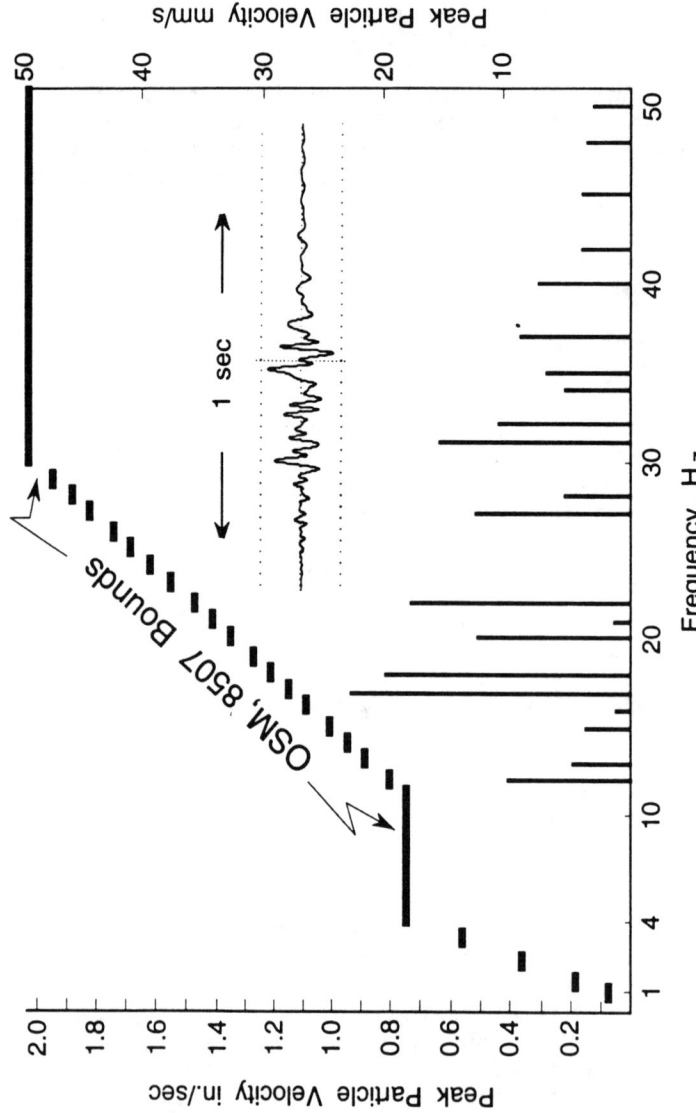

FIG. 7. Zero point crossing frequency analysis of the inserted particle velocity time history for all positive and negative excursions

Such graphical presentation also automatically facilitates comparison of allowable and actual particle velocities at frequencies between 10 and 30 Hz where the 8507 curve follows a constant displacement rather than constant velocity bound. This mid frequency range (or any frequency based criteria that follows a constant displacement bound) poses difficult control challenges. For instance, how high a particle velocity is allowed if its dominant frequency is 25 Hz? The 8507 bound is obviously greater than 0.75 ips (19 mm/s), but how much? Most field personnel have neither the time nor the inclination to calculate displacement from measured velocity and estimated dominant frequency. Unfortunately while such graphical comparison simplifies interpretation, it also requires that the time history be recorded digitally, a subsequent computerized analysis, and an additional graphical product for each time history so analyzed. These are the same extra steps necessary with Fourier frequency and SDOF response spectra calculations, which the visual zero point crossing simplification tried to avoid.

CASE STUDY OF VISUAL & SPECTRAL DETERMINATION OF DOMINANT FREQUENCY

Removal of some 200,000 cubic meters of rock in the urban setting shown in Fig. 8 involving over 50 structures within 60m (200 ft) of blasting involved use of the frequency dependent blast vibration control labeled "general urban" in Fig. 2. The plan view of the project (Fig. 8) shows the intertwined fingers of basalt and gabbro rock (identified by the "B & G" labels), required cuts of up to 12m (40 ft), and contour lines respectively. Relaxation of the 25 mm/s (1 ips) criteria for structures within 30 m (100 ft) of the blast when dominant excitation frequencies exceeded 40 Hz allowed efficient shot design while minimizing threshold cracking for all structures.

Early in the excavation process, it became obvious that there was no commonly accepted method of interpreting the time histories when the 25 mm/s criterion was exceeded within 30m (100 ft) of the blast. While monitoring instruments produced both a time history and a zero point crossing estimate of the frequency associated with the peak of the velocity time history, it was questioned whether slightly smaller

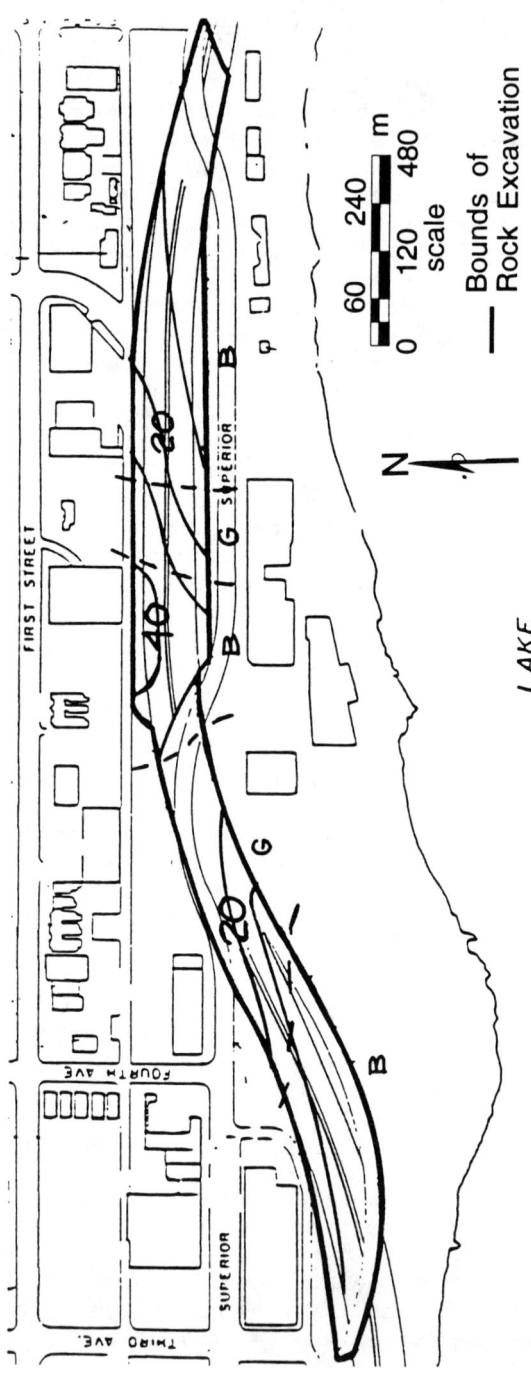

FIG. 8. Plan view of urban rock excavation and blasting for a highway in a downtown area

peaks could exceed the criteria. This concern lead to a study of some 17 time histories to compare visual and zero point crossing estimates of dominant frequency with response spectrum calculations (Linehan, 1985). The comparison is summarized in Tables 2 & 3, where the time histories are distinguished in Table 2 by shot & ID number, scaled distance (S. Dist.) and peak particle velocity (PPV) data and in Table 3 by ID number and relevant minimum and maxiumum frequencies.

While the original report presented time histories and 5% damped pseudo velocity response spectra from all 17 time histories, only those from cases 8,11,12,16,&17 are presented herein in Fig. 9. Blasts 31 & 35 (ID No.'s 11, 12 & 16, 17 respectively in Table 2) involved detonation of 6.3 and 6.0 kg per delay with only one delay per hole, 14 & 17 holes, and a total explosive weight of 77 & 91 kg, respectively. Shot 31 was arranged as a 3 row bench blast with a burden and spacing of 1.8m & 2.7m while 35 was arranged as a 2 row echelon with a staggered burden and spacing of 2.4m & 2.7m. Relevant peak particle velocities and scaled distances are given in Table 2. Shot number 31 with square, constrained shot geometry was dominated by unexpectedly high frequencies for its large distance to the blast monitor. Shot 35, with a long, thin shot geometry involved expectedly high dominant frequencies; however, the longitudinal component (17 L) involved uniformly lower than expected dominant frequency of 30 to 33 Hz.

TABLE 2
CASE STUDY PARTICLE VELOCITIES

Shot No.	ID No.	Dist (R) ft (m)	Wt (W) lb	S.DIST $R/W^{1/3}$	PPV ips	SI (mm/s)
28	8	140	12.8	39	0.63	
31	11 V	335	13.9	90	0.47	
"	12 L	335	13.9	90	0.40	
35	16 V	185	13.2	51	0.65	
"	17 L	185	13.2	51	0.90	

Generally, the time histories could be visually classified as two types: those typically characterized by a single or discrete frequency, and those containing a broad frequency range. Those records which were characterized by vibrations with a relatively narrow

TABLE 3 - FREQUENCY EVALUATION
VISUAL VERSUS RESPONSE SPECTRUM COMPARISON

ID No.	Visual time history evaluation (Hz)			Response spectra evaluation (Hz)			Pass	Fail	Frequency selection
	MIN.	MAX.	DISC.	MIN.	MAX.	PEAK			
1	40	100	X	45	110	70	X		40
2	25	100	X	55	150	70		X	X
3	X	X	35	30	40	35	X		35
4	20	30	X	20	85	60	X		20
5	25	35	X	22	35	25	X		25
6	50	85	X	45	110	100	X		50
7	X	X	85	80	150	100	X		85
8	60	>100	X	65	95	80	X		60
9	X	X	85	75	100	90	X		85
10	X	X	60	55	85	60	X		60
11	X	X	80-100	28	>150	150		X	X
12	25	60	X	25	95	30	X		25
13	35	100	X	29	70	60	X		35
14	X	X	57	42	63	50	X		57
15	X	X	65	65	75	70	X		65
16	25	100	X	60	85	70		X	X
17	X	X	33	20	42	30	X		33

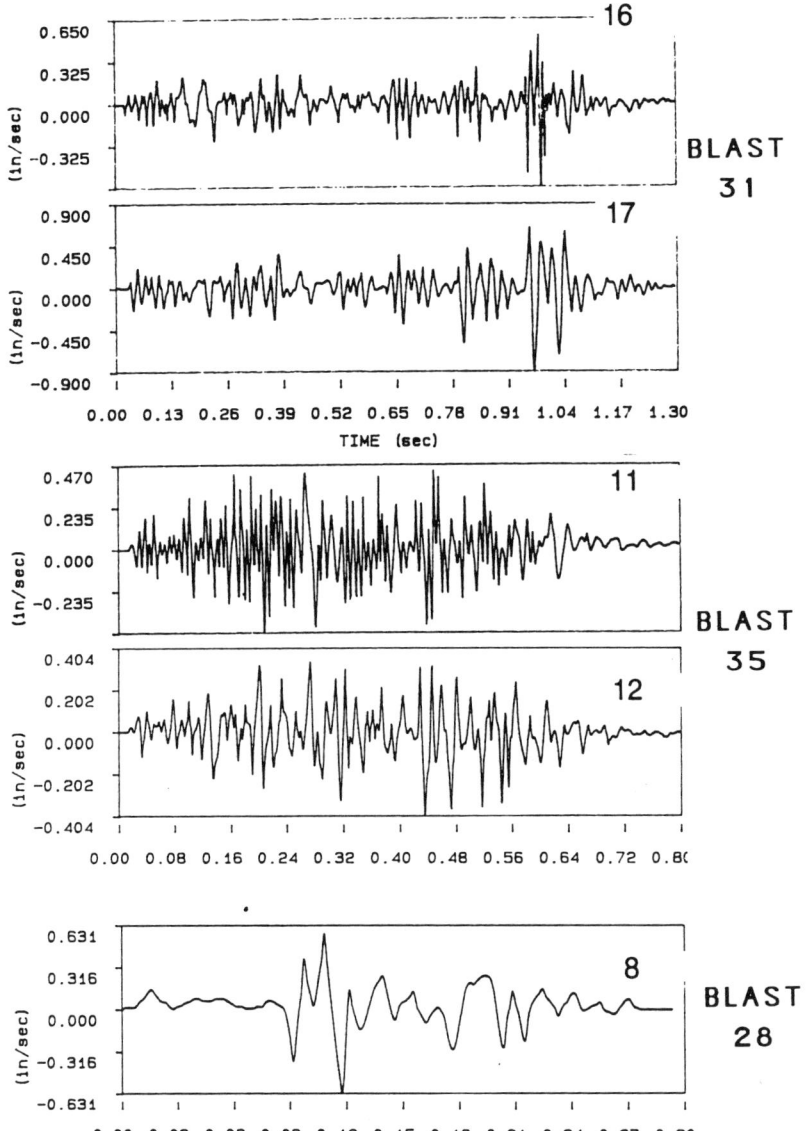

FIG. 9. Particle velocity time records from 5 of the 17 comparisons of visual determination and response spectrum calculation of dominant frequency

band of frequencies (ID's 3, 7, 9, 10, 11, 14, 15 and 17) have a representative frequency tabulated in the discrete frequency column, DISC in Table 3. A minimum and maximum frequency is given, in the columns identified as MIN and MAX respectively, for those records considered to have a broad band character.

Visually estimating the dominant frequency associated with time histories which had only one, or a "discrete" frequency was straightforward. Estimating the frequency range for those time histories containing a broad frequency range was difficult. Therefore, the following guidelines are proposed for visual evaluation.

Where the time history record typically shows one dominant frequency, the peak-to-peak time periods over several continuous cycles of vibration containing the peak particle velocity should be averaged. This method should be used only where time histories are obviously dominated by one frequency, such as Blast 35 (ID 17).

Where the blast time history shows a variety of dominant frequencies, as typified in several illustrated records (i.e., blast 28-ID 8) the lowest frequency associated with the dominant peak amplitudes should be employed. The peak amplitudes are those with amplitudes of at least 70% of the peak. This technique is a conservative approach and follows the method employed by the U.S. Bureau of Mines to establish the 8507 curve.

For comparison of the visual and computational approaches, 5% damped pseudo velocity response spectra were calculated and shown in Fig. 10. Calculated response spectra are characterized in Table 3 with three frequencies; the PEAK is the frequency of the system with the greatest response; the MIN and MAX were the frequencies that correspond to systems that had responses of 70% of that of the peak spectrum amplitude. If two peaks occurred in the response spectrum, the frequency of the lower peak is reported as long as the lower peak response is within 70% of the higher peak.

The 70% of the peak amplitude criteria for determining the MIN and MAX frequencies, i.e., $1/\sqrt{2}$, is an accepted measure of relative energy. The energy of

URBAN BLASTING CONTROL 203

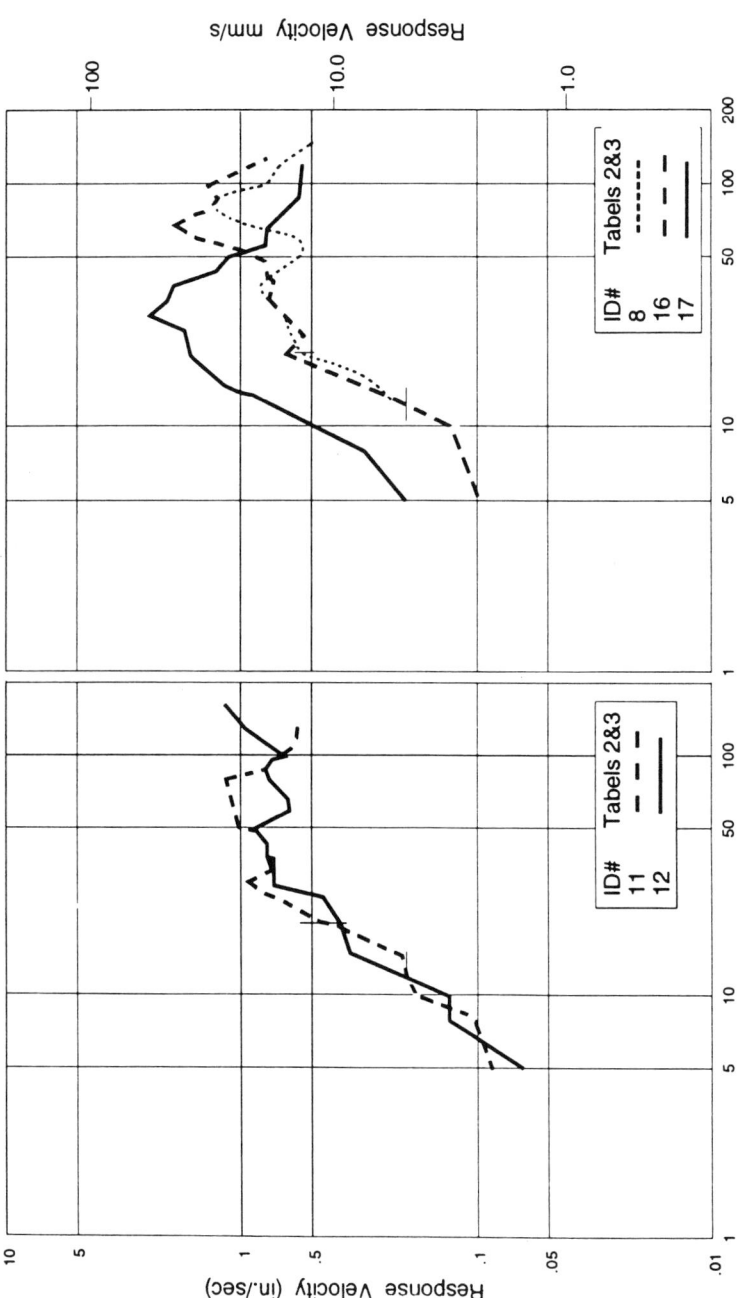

FIG. 10. Response spectra of 5 of the 17 comparisons of visual determination and response spectrum calculation of dominant frequency

a moving mass is proportional to velocity squared i.e.; $E = 1/2\ mv^2$. Therefore, a system moving with a peak velocity 70% that of another has 50% of the energy of the other: $(0.70v)^2 = 0.50v^2$.

This 70% criteria distinguishes frequencies that contain significant, at least 50%, energy over a broad range of frequencies from spectra that contain energy in a narrow range. For instance, the MIN and MAX values for ID 16 occurs over a relatively narrow frequency band of 60 Hz to 85 Hz. Other response spectra, such as ID 11 & 12 are characterized by a much broader dominant frequency range of 25 to 95 Hz.

As indicated in Table 3, there was acceptable agreement between visual estimates and response spectra calculations for 14 of the 17 time histories. Of the three time histories which were not satisfactory, two (ID 2 & 16) had a visually broad band character and would have been interpreted in an overly conservatively manner. The other (ID 11) was found by PVRS calculation to contain a significant low frequency component not revealed in the visual inspection, and would have been unconservatively interpreted. Thus, for 1 of the 17 complex records chosen for study was it necessary to calculate (as opposed to visually inspect) the dominant frequency.

CONSIDERATIONS FOR CLOSE-IN BLASTING WITH DOMINANT FREQUENCIES ABOVE 40 HZ

Dual monitoring of motions during close-in blasting for construction of an addition to Cornell University's Olin Library provides insight into the importance of high frequency motions (McKown, 1991; New, 1991; Thendean, 1992).

The basement for the additions to the library was excavated as shown in Fig. 11 between two other halls by blasting, which was controlled with the criteria shown in Fig. 12a. Extension of the high frequency bound in Fig. 12a is supported by response spectrum analysis (Dowding, 1972 & Hendron, 1979) and high frequency Swedish data (Langefors et al 1958). Most importantly response spectrum analysis shows that as long as dominant frequencies increase with peak particle velocity, relative displacement remains constant. Since relative displacement of walls

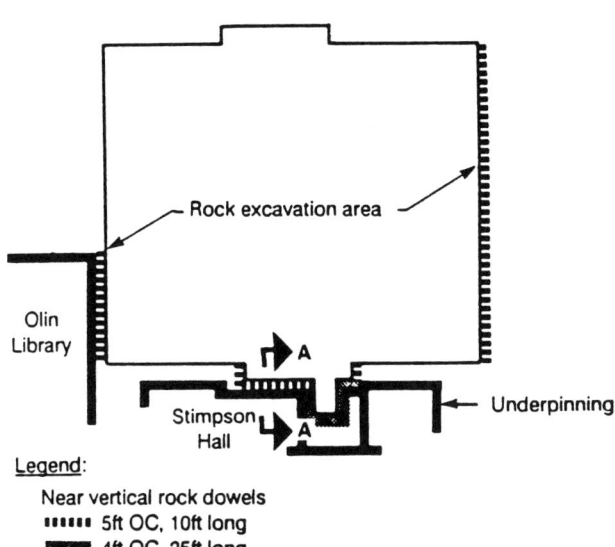

FIG. 11. Proximity of the Olin excavation to adjacent structures that shows the location of rock reinforcement employed to protect the rock against unplanned gas pressure induced movement (From McKown, 1991).

produces the strain that causes cracking, if the relative displacement remains constant, the probability of cracking remains constant. This physical explanation is substantiated by the high frequency data collected by Langefors and successful use of the constant displacement bound control on well supervised projects (Hendron, 1979; McKown, 1991; etc).

Excavation was monitored with conventional blast vibration instruments with their traditional 2 to 200 Hz range of flat system response. Thendean (1992) and New (1991) also monitored motions with another system

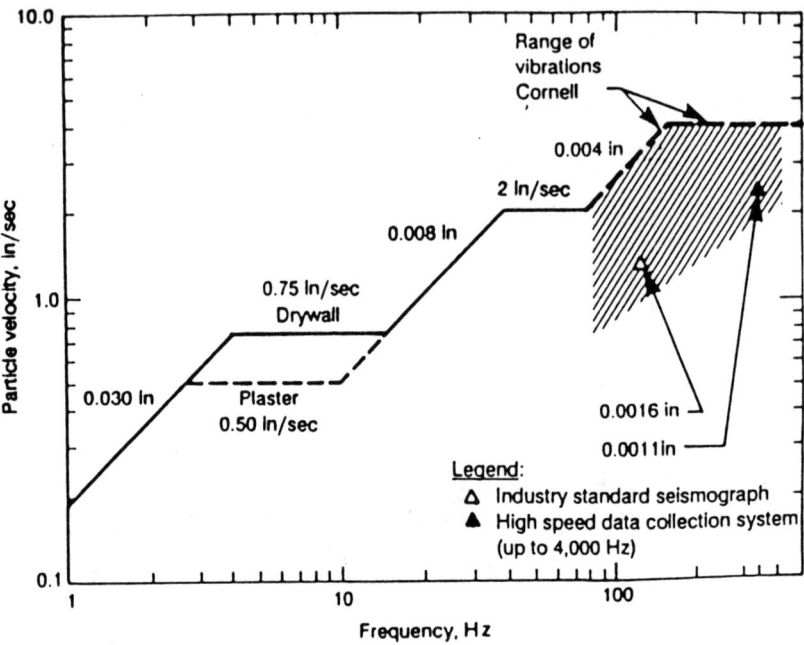

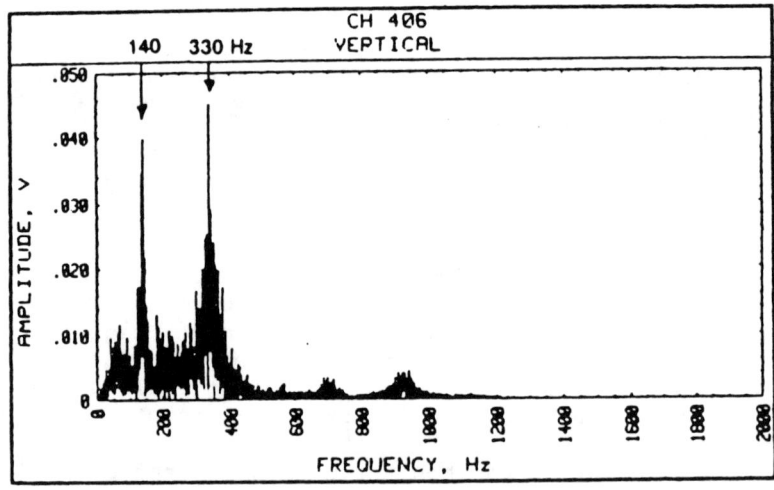

FIG. 12. Comparison of 8507 frequency based criteria and measurements a) and Fourier analysis of record obtained by high speed system b) (From McKown, 1991)

which employed velocity transducers and a digitization system with a flat frequency response that extended above 400 Hz. A comparison of peak motions measured with both systems shown in Fig. 12 shows that traditional systems may miss very high frequency motions. The Fourier frequency spectrum of motions recorded with the higher frequency system in Fig. 12 confirms the 140 and 330 Hz dominant frequencies which would result from monitoring with traditional and research systems, respectively.

Differences between the measurements shown in Fig. 12 raises the question, which system is most appropriate for monitoring? The principle that true phenomena should be measured would support a call for a new generation of high frequency vibration monitors. On the other hand, a focus on measurement of the greatest ground displacement or those motions likely to excite principal building components and equipment would support monitoring by both systems.

Observations in the 8507 curve shows that low frequency motions between 4 and 15 Hz are the most likely to cause threshold cracking in walls at low particle velocity levels. This range is that of the natural response frequencies of superstructure and structural components such as walls and floors (Dowding, 1985), where excitation is most likely to be amplified and result in large strains. Any machinery attached to these floors and walls is most likely to be excited at these low frequencies. Thus it appears monitoring typical structures and components with traditional or research equipment will not reduce the normal level of precaution. However, wherever response of machinery of structural components can be amplified at frequencies above 200 Hz, high frequency systems may be necessary.

OVERRIDING IMPORTANCE OF PERMANENT DISPLACEMENT WITH CLOSE-IN BLASTING

Discussion of very high frequencies should not detract attention from the most important blast effect to control when high frequencies are likely--permanent displacement. High particle velocities and high frequencies are most likely to occur within 30m (100 ft) of a structure. Within this distance cracking is more likely to result from permanent displacement of rock blocks beneath the protected structure, not vibratory shaking. The literature contains many reports of the effects of delayed gas pressure induced block movement (Dowding, 1985; McKown, 1991; Oriard,1991; etc.) and static slope instability that result from the excavation itself.

Permanent displacement becomes especially troublesome within 7.6m (25 ft) of a structure, because of the difficulty of blasting without exceeding particle velocity limits. This difficulty normally diverts engineering attention from stabilization of the remaining rock mass and concentrates it on vibration problems. Since vibration limits were developed from the observation of STATICALLY STABLE structures, they give no direction for dynamic or gas pressure induced instability of the remaining rock mass or for its deterioration by the propagation of explosive gases. As described by McKown and others, the remaining rock mass should be stabilized before blasting.

CONCLUSIONS AND RECOMMENDATIONS

This paper has developed the technical background for a frequency based particle velocity limit for the control of blasting operations. Details have been presented for three different methods to determine dominant vibration frequency that are necessary to control operations on the basis of frequency. Special attention was devoted to control of close-in construction blasts, which produce high particle velocities at high frequencies.

Within the framework of this discussion the following conclusions can be reached: 1) Current regulatory control limits are defined best for dominant frequencies between 4 and 20 Hz; 2) For dominant frequencies above 20 Hz, control is best exercised with a constant ground motion displacement bound no greater than 0.008 in: 3) A constant velocity limit for motions with dominant frequencies above 30 Hz was not experimentally verified by the US Bureau of Mines: 4) Permanent displacements may pose the most significant challenge for close-in blasting with high velocities/frequencies, and is many times neglected: 5) Visual techniques to determine dominant frequency such as zero point crossing are sufficient for the majority of blast monitoring: 6) When frequency calculations are necessary, Fourier or SDOF response spectrum methods are equally appropriate: and 7) Fourier and pseudo velocity response spectra are mathematically related.

The main issue for frequency based control of blasting is a guide for the necessity of calculating dominant frequency. The following approach, which reduces the need for excessive calculation, would involve incremental controls. First, a maximum allowable particle velocity would be established without regard to frequency, say 1 inch per second (or 25 mm/s). If 80% of this limit is reached, the contractors must produce a zero point crossing analysis while they describe in writing how the 80% limit will not be exceeded on the next blast. If the 25 mm/s limit is exceeded, drilling and blasting would be halted, a response spectrum analysis would be required and the shot initiation pattern modified to reduce the expected particle velocity on the next blast. Within a specified small distance, (say 30 m (100 ft)) of the blast, particle velocities greater than the general limit (say similar to that labeled U in Fig. 2) would be allowed if accompanied by a response spectrum analysis that is available before the holes for the next blast are drilled.

Future trends in blast vibration control in urban areas can be derived from the above conclusions and areas requiring further investigation can be ascertained from the limitations of the conclusions. Trends include increasing 1) field calculation of both Fourier and response spectra enabled by new generations of portable computers, 2) telemetric data acquisition enabled by new remotely operable cellular & radio systems in order shorten cycles of interpretation and 3) rock mass protection in high frequency/velocity situations.

Areas requiring further investigation include 1) Appropriate instrumentation schemes for removal of rock immediately adjacent to existing structures that account for the actual cracking mechanisms, 2) Methods for prediction of deleterious effects produced by excessive delayed gas pressure, and 3) Experimental verification of the velocity control limits in high frequency/velocity situations.

APPENDIX REFERENCES

Bendat, J.S., and Piersol, A.C (1980), <u>Engineering Applications of Correlation and Spectral Analysis</u>, John Wiley and Sons, New York, NY.

Crenwelge, O.E. (1988), "Method for Determining Amplitude-Frequency Components of Blast Induced Ground Vibrations," <u>Proceedings, 4th Symposium on Explosives and Blasting Research,</u> Society of Explosive Engineers, Montville, OH.

Crenwelge, O.E. (1991), "Transient Data Analysis Procedure for Reducing Blast-Induced Ground and House Vibrations," <u>Proceedings, 17th Conference on Explosives and Blasting Technique</u>, Society of Explosive Engineers, Montville, OH.

Dowding, C.H. (1971), "Response of Buildings to Ground Vibrations Resulting from Construction Blasting," Ph.D. thesis, Department of Civil Engineering, University of Illinois, Urbana, IL.

Dowding, C.H. (1985) <u>Blast Vibration Monitoring and Control</u>, Prentice-Hall, Englewood Cliffs, N.J.

Dowding, C.H. (1988), "Comparison of Environmental and Blast Induced Effects through Computerized Surveillance," <u>The Art and Science of Geotechnical Engineering at the Dawn of the 21st Century, R.B. Peck Honnorary Volume</u>, W.J. Hall, ed., Prentice-Hall, pp. 143-160.

DIN, (1983), <u>Deutsche Normen: Erschütterungen im Bauwesen - Einwirkungen auf bauliche Anlagen</u>, DIN 4150.

Fulthorpe, C.S. (1979), "Computer Modelling of Structural Response to Combined Air Blasts and Ground Vibrations," M.S. thesis, Department of Civil Engineering, Northwestern University, Evanston, IL.

Hendron, A.J. (1977), "Engineering of Rock Blasting on Civil Projects", <u>Structural & Geotechnical Mechanics</u>, W.J. Hall, ed., Prentice Hall, Englewood Cliffs, NJ

Hudson, D.E. (1979), <u>Reading and Interpreting Strong Motion Accelerograms</u>, Earthquake Engineering Research Institute, Berkeley, Calif.

Langefors, U., Westerberg, H., and Kihlström, B. (1958), "Ground Vibrations in Blasting," Water Power, September.

Linehan, P. W (1985), "Blast Vibration Time-History Evaluations," <u>Report No. 840659</u>, Wiss, Janney, Elstner Assoc., Northbrook, IL.

McKown, A. F. (1991), "Close-In Construction Blasting: Impacts and Mitigation Measures," *Proceedings, 17th Conference on Explosives and Blasting Technique*, Society of Explosive Engineers, Montville, OH.

New, B.J. (1991), Private Communication, Head, Tunneling and Geomechanics Section, Transportation and Road Research Laboratory, Cranthorne, UK.

Newmark, N.M., and Hall, W.J. (1982), *Earthquake Spectra and Design*, Earthquake Engineering Research Institute, Berkeley, Calif.

Oriard, L.L. (1991), "Close-In Blasting Effects on Structures and Materials," *Proceedings, 17th Conference on Explosives and Blasting Technique*, Society of Explosive Engineers, Montville, OH.

Siskind, D.E. (1987), "Private Communication," U.S. Bureau of Mines, Twin Cities Research Center, Minneapolis, MN.

Siskind, D.E., Stagg, M.S., Kopp, J.W., and Dowding, C.H. (1980), "Structure Response and Damage Produced by Airblast from Surface Blasting," *Report of Investigations 8507*, U.S. Bureau of Mines, Minneapolis, MN.

Thendean, G. (1992), "High Frequency Blast Vibration Analysis," M.S. Thesis, Dept. of Civil and Environmental Engineering, Cornell University, Ithaca, NY.

Veletsos, A.S. and Newmark, N.M. (1964), *Design Procedures for Shock Isolation Systems of Underground Protective Structures*, Vol III, "Response Spectra of Single-Degree-of-Freedom Elastic and Inelastic Systems", report prepared for the Air Force Weapons Laboratory, Kirtland, NM.

CONSTRUCTION INDUCED VIBRATION IN URBAN ENVIRONMENTS

Barry M New[1]

ABSTRACT

This paper reviews some of the work carried out at the United Kingdom Transport and Road Research Laboratory (TRRL) to investigate the effects of ground vibration caused by construction works.

The propagation, spectral distribution and measurement of ground vibration are discussed and data presentation using various scaling methods are described. Research to determine the potentially adverse effects of vibration on young concrete and reinforced soil structures is reported. The effect of decoupling explosives in oversized drill holes during presplit blasting is discussed. Extensive site investigations into the effects of variability in the delay elements of detonators are presented together with a numerical model which has been developed to improve prediction of vibration from multidelay blasting rounds.

Examples of vibration levels are given for a variety of construction sources (eg vibrating rollers, piling, tunnelling machines) which have been observed at sites in the UK.

INTRODUCTION

Throughout the world environmental considerations are increasingly playing an important role in the planning and realisation of construction projects. This paper considers the effects of vibration induced by a variety of construction processes and how the vibration caused may influence the works. The problems are generally most acute when construction takes place in densely populated areas and the financial consequences of rigid controls can constitute a significant burden on the project as a whole.

Vibration from construction works will generally be of a temporary nature, but the disturbance caused may result in substantial nuisance

---

[1]Head of Ground Properties and Underground Structures, Transport and Road Research Laboratory, Department of Transport, Crowthorne, Berkshire, RG11 6AU, UK.

to the local population and, in some cases, permanent damage to structures. Blasting and piling operations have in the past been the cause of greatest concern, but in recent years construction works have utilised larger equipment as economic pressures have forced greater emphasis on mechanised rather than labour intensive techniques. These developments have resulted in the use of machines that dissipate large amounts of energy, in the form of ground vibrations and noise, into the environment. Tunnelling works have followed this trend and, in particular, the use of full-face tunnelling machines in shallow urban works is now commonplace. Figure 1 illustrates the vibration levels caused by a variety of sources observed at construction sites in the UK and demonstrates the significance of piling operations and even the smallest of explosive excavation charges.

The market share for blasting in civil engineering construction has been reduced during recent years in favour of mechanised excavation techniques. This has arisen in part because the use of explosives, particularly in urban areas, is often considered undesirable on environmental, security or safety grounds. For construction works, in areas where vibration could cause problems, blasting is unlikely to be used unless the rock strength is so high that other methods are impracticable or uneconomic.

Predictions of vibration by analytical methods are often rather uncertain and in many instances they are not carried out at all. This lack of prior knowledge can lead to unexpected restrictions being placed on the Contractor during the actual works.

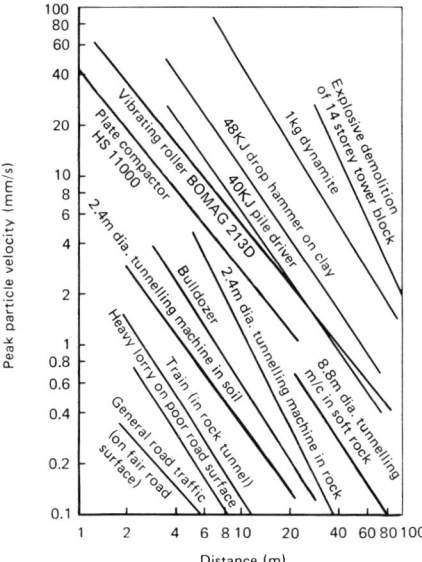

Fig. 1 Relative PPV from various sources

Most vibration associated problems may be expressed as two separate questions:

(i) what level of vibration will be produced by the proposed construction works? This will depend on the construction method and the seismic propagation characteristics of the site,

and (ii) what is the acceptable level of vibration? This will depend on the type of structures at risk and the sensitivity of the local population to nuisance.

The answers to both these questions are usually site specific, although an initial appraisal of hazard may be made which is based on experience from other sites. Reference to case history information will often assist planning operations. Where excavation by blasting is required in an urban environment it may well be prudent to carry out trial blasts as part of the site investigation programme.

'Acceptable' vibration levels vary considerably from country to country, and there is no internationally accepted standard defining them in relation to damage to structures. For example, German and Swiss standards are more conservative than the practice in the US, UK or Sweden (New, 1990a). A pragmatic basis for the more demanding standards does not appear in the literature and, despite considerable efforts, we have been unable to obtain a substantiated body of data which would support the general application of the more rigid standards. In the UK the British Standard 'BS7385 Measurement and evaluation of vibration in buildings' is soon to be supplemented by a second part 'Ground borne vibration - Guide to damage levels'. It is hoped that this document will provide guidance to help prevent damage to structures without being unnecessarily restrictive.

The effects of vibration on the environment during the construction and operation of railway tunnels is of particular current interest in the UK due to major railway construction in the South-East of England. The Channel Tunnel Rail Link is to connect the Channel Tunnel itself to Central London; it requires considerable lengths of tunnel beneath London and Ashford which are capable of carrying full gauge high speed trains. The Jubilee Line Extension connects the redeveloped London docklands to the London metro system and requires over 24km of bored tunnels plus station works. 'Crossrail' will provide a much needed east-west connection beneath the very heart of London whilst the Heathrow Express tunnels will provide a high speed route from stations beneath London Airport terminals to the railway main line into central London.

The ground conditions in SE England do not require the use of explosive excavation methods and therefore much of the effort has concentrated on the predication and mitigation of vibration caused by the operation of the railways (Crabb et al, 1991). Because the tunnels are often rather shallow and may pass beneath important buildings, the effect of re-radiated noise in the overlying structures is of particular interest. The choice of track design can reduce ground vibration caused by trains thereby lowering noise levels in overlying structures to acceptable levels; where this is not possible, route options may need to be reconsidered. Vibration caused by piling and other large equipment at work sites may also be a problem, although the ground movements caused by the excavation and tunnelling works generally dominate in these circumstances (New and O'Reilly, 1991).

This paper reviews procedures for the measurement and evaluation of vibration data and gives information on various applied research projects which have significantly benefitted both design and construction practice.

THE PROPAGATION, MEASUREMENT AND EFFECT OF GROUND VIBRATION

Propagation

The transfer of energy from construction sources to the ground mass and the subsequent propagation of the vibration involves very complex processes, particularly when the field conditions imposed by excavation works are considered. The near surface weathered rocks and soils, characterised by low in-situ stress conditions and with which most civil engineering works are concerned, present a particularly intractable set of problems.

Not only is the initial radiation pattern from such complex sources difficult to define, but it constantly changes with propagation through the ground mass due to the following factors:-

(i) geometrical spreading

(ii) the progressive separation of compressional, shear and surface wave types due to their differing propagation velocities

(iii) the presence of discontinuities in the seismic impedance which will cause reflection, refraction, diffraction, scattering and mode conversion processes

(iv) internal friction causing frequency dependant attenuation.

The combined effect of these factors causes the amplitude of the motions to be attenuated, particularly at high frequencies, and the wave packet duration to increase with propagation distance.

Where explosives are used in civil engineering works a pragmatic approach has prevailed and the use of various scaling methods based on field data are used for the prediction and evaluation of vibration data from explosive sources. These techniques are based on dimensional analysis and are widely used in various forms. The magnitude and spatial decay of blast induced vibrations have been discussed by numerous authors, although many of the data have come from the mining and quarrying industries, and care must be taken if this information is to be used in a civil engineering context.

There is, however, often a fundamental difference in the type of wave motion of concern during nearby construction in comparison with that normally associated with quarry workings. Quarrying will usually involve relatively large rounds of explosives at substantial distances from residential structures. It is therefore usual to measure these vibrations at many tens or hundreds of metres from the source. At these distances, and with a near-surface source, the predominant ground vibrations will usually be due to surface wave motions at low frequencies. In contrast, construction induced vibrations may be induced by small charges at very close range. In these circumstances body wave motions will dominate and much higher frequencies must be considered.

Measurement

Harmonic vibrations may be described by any two of the following parameters: frequency, peak particle displacement, peak particle velocity (PPV) and acceleration. With regard to human perception of vibration, the relevant parameter is related to the frequency range involved. At frequencies below 10 Hz acceleration seems to be the dominant consideration, whereas at higher frequencies velocity or displacement criteria may be appropriate.

Transducers are readily available with output voltages proportional to particle velocity. Geophone types are self-generating (need no power supply) and are ruggedly designed for field use. Measurement of particle velocity allows single-process derivation of acceleration, by differentiation, and of displacement by integration if required.

Measurement of particle velocity should be made in three mutually perpendicular directions. This allows the calculation of the resultant particle velocity, V_R, by the vector summation of the component velocities V_1, V_2, and V_3:

$$V_R = (V_1^2 + V_2^2 + V_3^2)^{\frac{1}{2}}$$

Vibration measuring instruments that display the resultant peak particle velocity directly on a chart record or in digital format are available. If such an instrument is not used, the three vectors must be summed by 'hand' or computer processing. The true resultant is calculated by summing the three component values at simultaneous times. The 'pseudo resultant', sometimes referred to, is obtained by summing the maximum value obtained for each component during the period of the vibrations.

It is vital for the specification of the instrument chosen to be appropriate to the vibrations that it is to record, particularly in terms of frequency response and sensitivity. For instance, some instruments are limited to an upper frequency band of 200 Hz. This type of equipment may generally be quite satisfactory but will, of course, be insensitive to vibrations above 200 Hz that could be present. High frequencies (200 - 1000 Hz) will often be encountered in the region close to construction works. It has been common

CONSTRUCTION INDUCED VIBRATION

practice both in the US and UK to specify a limiting value for peak particle velocity at a site and then control the works with equipment incapable of measuring particle velocities at the high frequencies which may be present. This is an unacceptable practice which is not excused by vague assurances that 'high frequencies are not damaging to the structures' because in some circumstances it can be the higher frequencies and the associated enhanced accelerations which can be of particular significance to the serviceability or use of a building (e.g. electrical switchgear, computer installations, hospital operating theatre). Also, for structural components that follow the impressed ground vibrations (foundations, etc) the potentially damaging strains imposed are likely to be proportional to the particle velocity <u>whatever</u> the frequency.

Rocks and soils tend to act as low pass filters, that is, low-frequency vibrations are subject to less attenuation during propagation than higher frequencies. Figure 2 shows how the energy spectral density from a 1 kg dynamite charge (fired in a strong psammitic rock) varied when measured at various ranges. Note how, close to the charge, the maximum energy is in the 200-300 Hz band and significant energy is present at up to 1 kHz. This situation changes rapidly away from the near field, and at 23 m the energy peak occurs at about 100 Hz. At 140 m, the energy is contained in a relatively narrow band between 20 Hz and 60 Hz. Thus, equipment suitable for measurements at ranges of 25 m and beyond may not, in this particular case, be suitable for use close to the source.

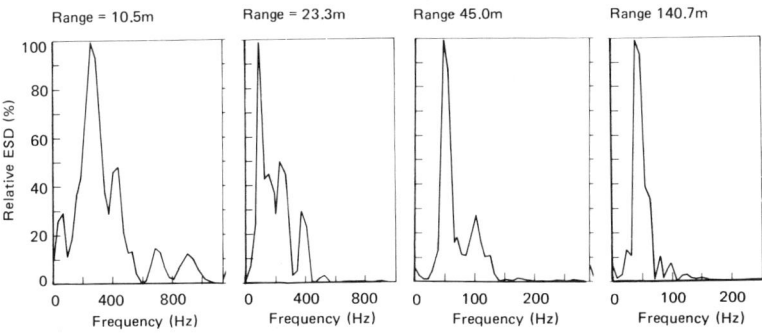

Fig. 2 Effect of source distance on energy spectral density (1kgf charge)

"Sandbagging"

Figure 3 shows a typical vibration/sound measuring instrument in use at a blast construction site. Note that the vibration transducers are being 'held' down to the ground by a sand bag placed on top of them. Because these transducers were close to the blasts the Engineer was convinced that a sandbag was necessary to keep the transducer in contact with the bedrock. As the blasting came even closer to the transducers more sandbags were added to cope with the higher vibration levels! This practice is, of course, totally misguided because where ground accelerations begin to approach 1 g no amount of sandbags will keep the transducers in contact with the ground. In these circumstances the sensing transducers <u>must</u> be firmly fixed with screws or adhesives to the rock or structure. Only then will they move with the ground and provide a true measure of vibration.

For vertical ground motions the transducers will leave contact with the ground at 1 g and horizontal motions greater than about 0.2 g may cause transducers to slip sideways. In our experience Figure 4 may be used to assess when (in terms of PPV, and estimated frequency) transducers must be rigidly fixed to ground or structure. For construction induced vibration, where the transducers are located relatively close to the source, it will usually be necessary to fix the transducers rigidly when particle velocities are expected to exceed a few millimetres per second.

Recently developed vibration monitoring equipment is frequently used just to print out the maximum resultant particle velocity due to a blasting event. Whilst this practice may be acceptable for routine works, occasional examination of the time history traces (particle velocity versus time) must be carried out by an experienced engineer capable of recognising symptoms indicating equipment or system faults (such as loose transducer arrays or signal saturation). This may avoid considerable difficulties where problems occur and the time histories are subsequently examined in detail during the consideration of a claim.

Fig. 3 Vibration measuring equipment

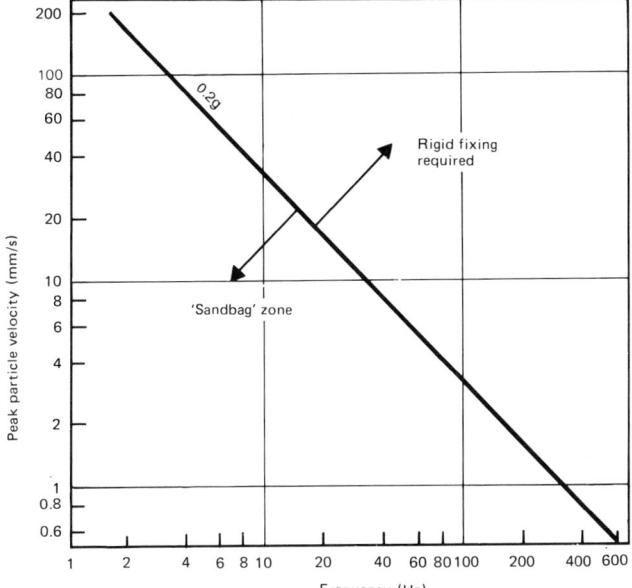

Fig. 4 Transducer fixing zones

Effects on structures

It is most useful for peak particle velocity (PPV) to be measured as it has been found to be best correlated with case history data of damage occurrence and has a theoretical underpinning in as much as the strain induced in the ground is proportional to the particle velocity. Although PPV is widely used to quantify the damaging potential of a vibration it must be recognised that 'velocity', of itself, cannot induce damaging forces. Such forces are generated in structures by both:-

(a) differential displacements which give rise to distortion as the structure follows movement of the ground upon which it is founded or

(b) change in the ground particle velocity vector which produces inertial forces upon the structure.

In practice the structure will be subjected to both 'distortion' and 'inertial' mechanisms at the same time and these will be superimposed upon pre-existing stresses and strains from other causes; damage will occur when the combined effects exceed the tolerance of the structure. Vibration can also give rise to longer term ground movements (e.g. by compaction) which may also contribute to structural distress.

For convenience Figure 5 separates the distortional and inertial factors and considers particle motions normal and parallel to the surface. The ground distortion (Figures 5a and b) could be attributed to vertically polarised shear waves of the vertical component of Rayleigh surface waves whilst the dilatation in this case (Figures 5c and d) results from a compressional wave type.

It can be shown that the shear strain (γ) imposed by the ground distortion at foundation level is a function of the particle velocity (V) and the velocity of (shear) wave propagation C_s. That is $\gamma = V/C_s$. Similarly the strain (ϵ) caused by dilatation depends on the particle velocity and the compressional wave velocity C_p. That is $\epsilon = V/C_p$. This form of calculation will be appropriate for any structure which closely follows the movement of the ground, e.g., pipes, tunnels, foundations.

For most buildings however the actual distortions will be heavily dependent on the dynamic response of the structure. The natural frequencies and damping characteristics of a building will determine the strains imposed by the ground motions. To simplify, the case is considered where the ground motion wavelength is long compared to the length of the structure.

A force is required at the interface of structure and ground to move the building either up and down or side to side, Figures 5b and d. These forces will depend on the effective mass of the structure and the acceleration imposed by the ground wave motions. The acceleration (a) is a function of particle velocity (V) and the frequency (f) of

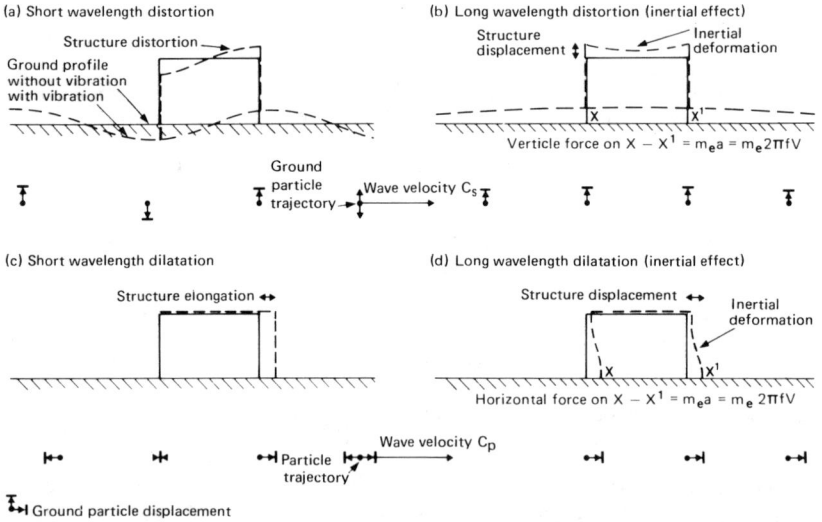

Fig. 5 Deformation and inertial forces due to ground wave motion

the motion (a = 2πfV for harmonic motion). The actual forces transmitted within the structure are dependant on its particular response characteristics but for a given structure the ground motion parameters concerning damage hazard may be taken as particle velocity and frequency.

In practice structural damage may result from the complex interaction of the mechanisms shown and it is important to emphasise that, whilst PPV may be the single most valuable parameter to observe, the frequency and propagation velocity of the ground motions must also be considered. A more detailed discussion of structural response using a single degree of freedom model is given by Dowding (1985).

Although it is essential to recognise that the risk of structural damage is also dependent on frequency and propagation velocity it seems likely that vibration damage criteria will continue to be related empirically with peak particle velocities. An important pragmatic influence on this approach is that there is little reliable information available on the damage induced in structures related to measured dynamic stresses and strains. The data that are available on distortions of a quasi-static nature are not readily interpreted in terms of dynamic movements because the effective strength of a material is often critically dependent on the rate at which the load is applied.

Effects on unstable ground

Besides damage to man-made structures, consideration must sometimes be given to potentially unstable soil or rock conditions in the vicinity of construction works. Cohesive soils are unlikely to be adversely affected by vibration, whereas loose sands may be caused to consolidate causing settlement. In exceptional cases liquefaction may take place and result in large ground settlements.

Blasting vibrations may also impose substantial dynamic loading on nearby rock slopes, which may result in rock fall. This problem can occur in tunnel portal areas where slopes are often at their steepest and the blasting is close to the surface. By the nature of their formation such slopes may have static factors of safety approaching unity and any additional dynamic loads may result in failure. An example of this type of failure is shown in Fig. 6. During the initial blasting work for this tunnel some 3000 tons of rock was dislodged from the daylighting discontinuity (a rock joint or flow banding plane), which is clearly seen just above the portal stonework. Incidents of this kind can prove hazardous to the workforce and may require costly remedial actions.

The analysis of rock slope stability has received considerable attention, but most texts tend to consider only the 'static' (gravitational, hydrostatic, etc) loadings that affect stability. Although it is reasonably straightforward to calculate the imposed dynamic loads (elastic and inertial) due to a given vibration, it is in practice difficult to assess their effects on the overall stability of the slope. Allowance for even quite moderate values of PPV will result in major destabilizing dynamic stresses in the stability equations. Calculations of this kind often give very pessimistic values for the slope's 'factor of safety' that are not evidenced in the field.

Fig. 6 Tunnel portal showing slope failure

THE PROCESSING AND PRESENTATION OF BLASTING DATA

Square and cube root scaling methods

Broadly speaking, the ground motions resulting from a blast will depend on the weight of explosive fired, the distance between the explosion and the observation point, and the ground characteristics. The effect of each of these factors is complex and no satisfactory theoretical approach for calculating the form of these motions has been developed to date. Therefore, scaling of field measurements is used almost exclusively to predict the magnitude and character of the vibrations from explosions.

A wide range of field data is available and several similar empirical approaches are in common use. The principal variables are usually related by an equation of the form:

$$PPV = KM^\alpha r^{-\beta} \qquad (1)$$

where PPV is the peak particle velocity, M is the charge weight and r is the distance from the explosion. The constants K, α and β are dependent on conditions imposed by the site and the type of explosion.

Two special cases of this formulation are most commonly used:

$$PPV = K(r/\sqrt{M})^{-n} \text{(square root scaling)} \qquad (2)$$

and

$$PPV = K(r/\sqrt[3]{M})^{-n} \text{(cube root scaling)} \qquad (3)$$

where, again, n is an empirical constant. Values of K and n typically range from 700 to 2000 and 1.5 to 2.0, respectively (for M in kilograms and r in meters).

The "square root" and "cube root" scaling methods both allow simple graphical presentation of the derived site laws. These laws are shown graphically (see Fig. 7) as a straight line (on a log-log plot) with a slope of -n and an intercept of K at unit-scaled distance.

Because of the difference in the forced scaling of r to M, these two methods lead to different predictions of peak particle velocity based on the same field measurements. Where extrapolation beyond the bounds of the field data is not required, the predictions will often be rather similar regardless of which method is used. However, where prediction is required beyond the range of charge weights or distances covered by the trial blasts, significant differences between the two methods may result. This situation is clearly unsatisfactory, as is the somewhat arbitrary scaling inherent in both methods.

It has been argued that these empirical laws should be "shaped" by dimensional analysis (Ambraseys and Hendron 1968; Newmark 1968). The dimensional analysis results in "cube root" scaling laws for explosions of different sizes in the same medium. This approach has led to equation 3 being used extensively.

It has been suggested (Ambraseys and Hendron 1968) that square root scaling has "no basis in dimensional analysis" and that cube root scaling should be used if estimates of motions are required that need extrapolation beyond the existing data. However, experimental evidence indicates that in certain situations, "square root" scaling does normalize the data very well. That is, the correlation coefficients were higher using $r/\sqrt{M}$ scaling than for $/\sqrt[3]{M}$ scaling.

In fact, because site conditions rarely comply with the assumptions made for the dimensional analysis, neither scaling method is strictly appropriate, and the best estimate of relative scaling between r and M is site and method specific.

It is usual to present blast vibration data in scaled distance graphical format with a "best fit" straight line obtained by linear regression analysis. A great volume of data in this form is available in the literature and, almost without exception, the peak particle velocity is well represented by a power law decay with scaled travel distance. That is, the measured PPV decay can be represented by a straight line with negative slope on a log-log plot, although the actual slope and intercept values may vary considerably from site to site and with different blasting conditions.

More complex forms of equation have been proposed in an attempt to include frictional dissipation effects. Such equations generally take the form

$$PPV = K(r/\sqrt{M})^{-n} \exp(-ar) \qquad (4)$$

where a is the spatial attenuation coefficient.

This type of equation allows a nonlinear regression line to be fitted to the data, thereby improving its correlation coefficient. However, because the majority of site data are well fitted by linear equations (on log-log plots), the additional complication is unlikely to find wide application. Moreover, this formulation is only appropriate at a single harmonic frequency, which is an unacceptable assumption for construction sites that are particularly close to the source (see Fig.2). Also, calculations based on the known frictional properties of rocks indicate that losses from this particular mechanism are unlikely to be significant at ranges of interest from most construction sources.

Site-specific scaling using multiple regression analysis

After having obtained field data relating peak particle velocity to explosive charge weight and range, it is clearly important to process and present the information in the best and most useful manner. Although the forced scaling of r to M implicit in equations 2 and 3 may be unsatisfactory to some extent, it does allow simple graphical presentation of the data. However, the exploitation of equation 1 allows site-specific scaling of r to M, which will enable predictions to be made using equations better correlated with field data. The improved correlation is the direct consequence of the more versatile equation with three variables (PPV, M and r), rather than with two variables (PPV and "scaled distance", $r/\sqrt{M}$).

The constants K, α and β in equation 1 may be determined by transforming the equation and applying a three-variable multiple linear regression analysis to the data as follows:

from eq 1:-

$$\log PPV = \log K + \alpha \log M - \beta \log r \qquad (5)$$

In this form, "best" values for K, α and β may be calculated based on the usual regression criteria that the sums of the squares of the deviations shall be minimized. For simple regression (two variables), these deviations are taken as deviations from a straight line, whereas for this three-variable analysis they are represented by deviations from the plane KABC, shown in Fig. 8.

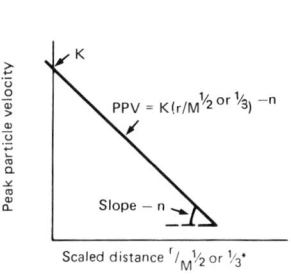 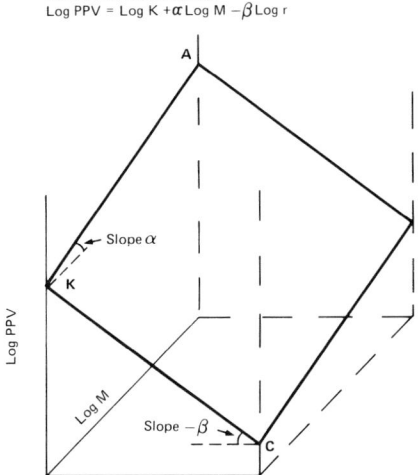

Fig.7 Scaled distance site law format   Fig. 8 3-D site law format

Although the projection shown in Figure 8 is useful in visualizing the equations 5 or 6, it is not appropriate for routine graphical presentation of data. It is possible, however, to plot the data, using the constant and exponents calculated by the multiple regression, in the convenient PPV v "scaled distance" format. By transforming equation 1 to the form

$$PPV = K(r/M^{\alpha/\beta})^{-\beta} \qquad (6)$$

(where the term $r/M^{\alpha/\beta}$ is the "scaled distance") the data may be presented in the same convenient format as for square or cube root scaling (see Fig.7), but without the forced scaling relationship. General comparisons of varying site law regressions may require reduction of the data to a unified scaled distance.

The coefficient of determination for the three variable regression will always be better than that obtained using the two-variable analysis for square or cube root scaling. (Except where, by chance, α/β is equal to 1/2 or 1/3, in which case the correlation will, of course, be numerically the same). Routine statistical tests of the significance of the correlations can be carried out and confidence limits calculated for the predictive equation if required.

The benefits of 'site specific' scaling can be considerable when site conditions or methods result in a scaling of r to M which substantially differs from the 'cube' or 'square' root assumption. Such a case is described below.

EFFECTS OF DETONATOR VARIABILITY

Field measurements

The comparison of vibration predictions (based on site investigation blasting trials) and the vibration measured during the construction of a major road tunnel (New, 1989) has indicated that normal scaled-distance site law analyses of trial data can overestimate the actual vibration caused by the works. Extensive vibration measurements were taken between mid 1986 and late 1987 during the construction of the Penmaenbach tunnel. Geophone and accelerometer transducer arrays were located in an adjacent railway tunnel, within the new tunnel and on surface rock exposures. Blast to transducer range varied between 4 m and 250 m and charge weights per delay between 2.35 kg and 54 kg. The effect of detonator inaccuracy was particularly notable during the firing of the longer duration delays (½ second delay series) firing the heaviest charges in the tunnelling round. This research has identified the cause of the discrepancy and provides techniques which now allow more accurate predictions to be made.

To explain the pattern of vibration produced by tunnelling works, it is necessary to describe briefly the sequence of a typical 'pull' which advances the heading into the rock. Figure 9 shows a sketch of a drilling pattern for the excavation of a tunnel with a face area of about 100 m². This type of round is known as the 'burn cut' and has become increasingly common since the introduction of multi-boom drilling jumbos, which are at their most efficient when drilling large arrays of parallel holes. The explosives are initiated from the centre outward as follows:-

> A series of short delay (millisecond series) detonators sequentially fire the holes numbered 1 through 9 (inside circles) using the larger uncharged holes, marked R, to provide some relief for bulking of the central pulverised core. The later delays numbered 3 through to 12 then fire in generally semicircular groups progressively away from the centre exploiting the partial void formed by the explosion of the earlier charges. After the bulk of the rock has been fragmented, mainly at ½ sec delay intervals, the perimeter holes, which are usually more closely spaced and lightly charged, are fired to produce a smooth finish with minimum overbreak and damage to the remaining rock. Finally, the 'lifting' charges are fired; these excavate the bottom of the heading and lift the rock fragmented by the main part of the round back from the face. This loosens and spreads the rock pile making mucking out easier.

Efficient blast design which ensures both good fragmentation and economical use of explosives will exploit the full range of delays available, and tunnelling engineers have come to rely on the use of long delay detonators. The number of holes fired with detonators of a single nominal delay period will vary according to the blast design.

CONSTRUCTION INDUCED VIBRATION

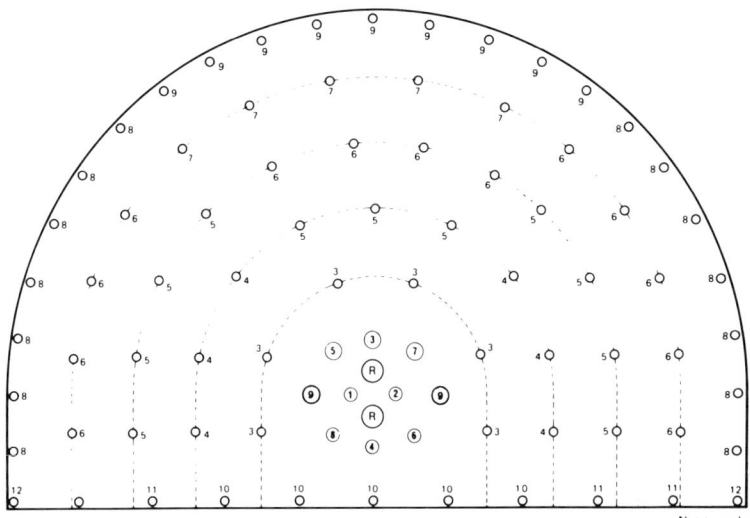

Fig. 9 Typical drilling pattern for 'Burn cut' tunnel round

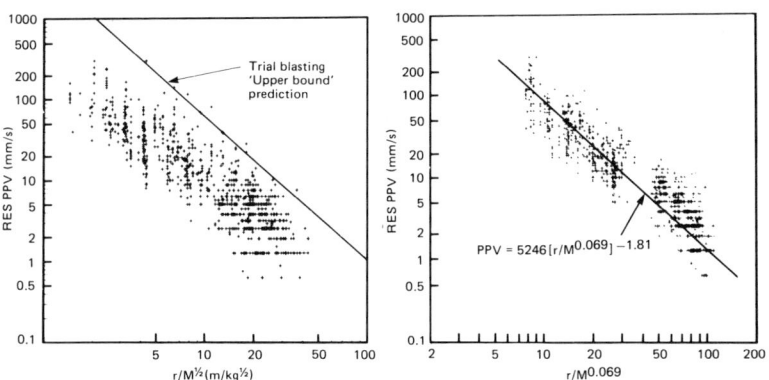

Fig. 10 Data from tunnel blasting (square root scaling method)

Fig. 11 PPV v Site specific scaling (from multiple regression analysis)

The early delays fire only a single hole, whereas the later bulk removal delays may fire over twenty separately initiated and charged holes of the same nominal delay period.

The peak particle velocity was initially plotted against square root scaled distance (Figure 10), as had been found most satisfactory for the presentation of the trial blasting data. It quickly became apparent that the actual PPV's were generally lower and much more widely scattered than the trial blasting had predicted. It is of interest to note however that the 'upper bound' equation did provide an excellent prediction of the maximum PPV's actually observed. The variation of PPV with range, r, and charge weight, M, were then examined individually. Whilst the PPV's were very clearly dependent on range no such well defined relationship was apparent for charge weight. A multiple regression analysis confirmed these observations and the following relationship was calculated:

$$PPV = 5246 \ r^{-1.81} . M^{0.125} = 5246 \ (r/M^{0.069})^{-1.81} \tag{7}$$

This relationship has been converted to site specific scaling and presented in Figure 11. Whilst the PPV of the vibrations depended on range, as predicted, it was only weakly dependant on charge weight per delay, as shown by the exponent value for M of only 0.125. $[(1/M^{0.069})^{(1.81)} = M^{0.125}]$.

Time history printouts comprised about one second of very complex motions, as the short duration delays fired in and around the centre of the face, followed by groups of wave packets from each successive delay until the vibration from the last delay (6 secs) had died away: the wave packet having a duration of about 6.5 seconds. Inspection of the records revealed that each group of vibrations, attributable to a nominal delay, comprised a number of separate individual detonations.

Well over a thousand individual hole firing times were identified from 25 separate tunnel rounds and the standard deviation for each period of long delay detonators was calculated. The standard deviations observed were similar to those given in the manufacturers specification and are typical of those associated with both electric and nonelectric detonator types.

It is of interest to note that, had the detonators all fired precisely at their nominal delay period, then vibration levels in the adjacent railway tunnel would probably have exceeded the prescribed limits. The variations of the delay period effectively spread the release of explosive energy thereby substantially reducing the amplitude of the vibrations to well within the contract limits. The introduction of new electronic delay detonators which are claimed to be extremely precise could, apparently, increase vibration levels, particularly on long delays. However, the effect should be mitigated by the use of many delay periods rather than the ½ second series currently available from detonators using pyrotechnic delay elements. Indeed, the use of detonators which may fire a prolonged series of relatively closely spaced delays would permit the individual charge weight to be kept low. Besides reducing peak particle velocities this would also shift the frequency of the motions to higher frequencies where most structures are considerably less sensitive.

Care must also be taken to avoid detonator delay periods that may excite natural frequencies in nearby structures. For instance, a series of detonations at 25 ms intervals may cause large motions in a structure with a natural frequency of 40 Hz (1/25 ms). It is usually relatively straightforward to design blasting rounds so that this does not occur.

The numerical model

The vibration levels observed during the Penmaenbach tunnel construction were generally lower than those predicted, because the predictive method assumed that all the charged holes on a nominally identical delay were fired instantaneously: The particle velocity time histories showed that this was not the case.

A numerical model was developed to estimate how the PPV might be expected to increase with the increase in charge weight per delay. It must be noted that, for a tunnel round of the type used at Penmaenbach, the charge weight per delay is increased by firing a larger number of similarly charged holes, not by increasing the charge weight in each hole. In these circumstances, the PPV would be expected to increase as the square root of the charge weight if the firings were at precisely the same instant. (note that for a single spherical charge the PPV would be expected to increase as the cube root of the charge weight).

A computer programme (SIMWAVE) was written to assemble the particle velocity time history of a wave packet comprising a number (N) of superimposed single hole detonations fired with timing errors normally distributed about a nominal delay period.

SIMWAVE requests the following inputs:-

1. The number (N) of holes fired on the delay
2. The standard deviation of the detonator delay period and the nominal delay period
3. A data file containing the particle velocity time history from the firing of a single hole. This can be a waveform recorded in the field, or a simulated waveform with specified frequency and damping characteristics.
4. The required number of repetitions of the calculations. (Because of the probabilistic nature for the computation it is necessary to obtain a measure of the variability of the predicted PPV's for a number of wave packet simulations).

SIMWAVE then performs the following operations:-

1. Randomly selects N initiation times with the specified mean and standard deviation.

2. Loads the appropriate source waveform from a data file or generates an artificial waveform as required.
3. Stacks the N individual waveforms to yield the complete wave packet time history.
4. Stores the maximum PPV occurring in the wave packet.
5. Repeats processes 1 - 4 for the specified number of wave packets to be generated.
6. Calculates the mean, maximum and standard deviation of the maximum PPV values for each wave packet.

The model was tested using a number of different single hole source waveforms recorded at the site. The output was found to be relatively insensitive to changes in waveform and yielded similar results for waveforms with similar PPV's, duration and frequency content.

SIMWAVE was run for a range of values (N = 1 to 30) for the number of individual holes fired on a single delay and the results are plotted in Figure 12. At Penmaenbach the maximum charge weight per delay was about 54 kg comprising 23 holes each charged with 2.35 kg of 80% strength special gelatine (SG80). Usually this maximum charge was fired on the 4 or 4.5 second delay. The standard deviation of the detonators was taken as 0.1 seconds which is typical of the longer delay detonators. Figure 12 indicates how the PPV (normalised with respect to PPV from a single hole firing) caused by a given delay depends on the weight of charge fired (M) which is proportional to the number of holes fired (N).

A best fit straight line has been plotted in that region of the curve which most closely applies to the field blast design. That is, the range of numbers of holes fired on a single delay was 5 to 30. The regression line indicates that, in this region, the PPV increases as $N^{0.14}$. This compares well with equation (7) which shows that, during the construction blasting, the PPV increased as $M^{0.125}$ $[(1/M^{0.069})^{-1.81}]$. Figure 12 also shows how the PPV might be expected to increase with charge weight when all charges are fired precisely at the same moment.

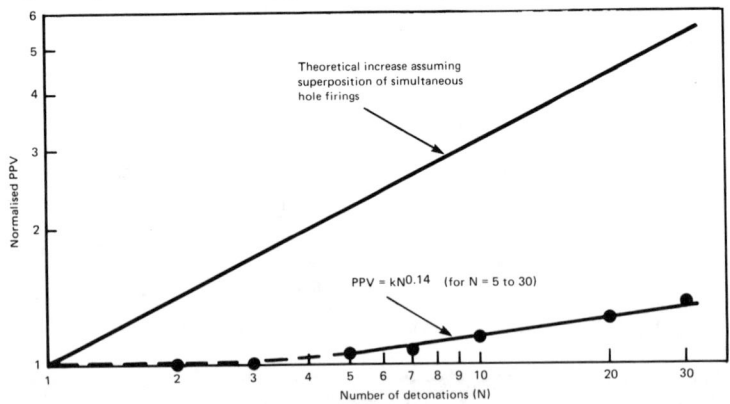

Fig. 12 Increase in PPV (normalised) with increase in number of holes fired

The numerical model provides predictions which are closely correlated with the field measurements at Penmaenbach. The model is being used to assist with the prediction of ground vibration at other sites, and this work, together with further production blasting measurements, will serve to test its validity under a variety of conditions.

VIBRATION DAMAGE TO YOUNG CONCRETE

Vibration caused by construction works may cause damage not only to pre-existing structures but also to the new work in progress. For example, there is often a need to excavate rock by blasting close to recently poured concrete. There is very little reliable experimental data which relates damage to limits of vibration or dynamic strain and some existing practice is suspected of being overly conservative. The establishment of rational controls has therefore been considered of importance so that costly constraints are not unnecessarily imposed on contractors methods.

Field experience indicates that mature concrete is extremely resistant to vibration and little evidence exists that young concrete is particularly vulnerable either. Laboratory trials have sometimes been open to criticism in that the dynamic strain in the test samples has not been well interpreted in terms of the particle velocity of the vibration. That is, confusion has occurred between the free body motions of samples subject to impacts and the critical and complex strain wave propagating within the samples.

A series of laboratory tests was carried out at TRRL to determine the onset of damage in young concrete. The laboratory trials used rectangular concrete test prisms subjected to impact induced strains which were measured directly by special resistance strain gauges embedded within the samples. A variety of concrete mixes were cast and subject to impact induced strains at ages between 11.5 and 45 hours. The impacts were steadily increased until cracking of the concrete had occurred. In all cases it was found that the cracking of the concrete was first observed by a distinct change in the form of the strain gauge output traces. This change was caused by the new discontinuities in the sample dissipating the wave motions and preventing free passage and end reflection for the induced motions. The next impact always caused cracks to become visible on the whitewashed sample surfaces.

Figure 13 relates tensile failure strain to cube crushing strength. Although the cube strength depended on the mix and age of the pour the dynamic tensile strain at failure was relatively independent of variation in age or mix. Failure of all samples tested occurred at tensile strains between $70\mu S$ and $130\mu S$. These observations are rather similar to those for mature concrete under static loading where tensile strains of between $100\mu S$ and $200\mu S$ are considered to be potentially damaging.

With a knowledge of the wave velocity and Poisson's ratio for the samples the measured failure strains have been interpreted in terms of equivalent particle velocity. These calculations indicate that the samples were undamaged by vibration with a particle velocity of up to 200 mm/s. These results may be applicable to normal mass concrete in intimate contact with the ground, such as foundations.

Where the structural concrete (columns, beams etc) is potentially at risk 'particle velocity' will not be an appropriate damage control criteria. In these circumstances it may be appropriate to set a maximum allowable tensile strain based on $70\mu S$ reduced by an adequate factor of safety.

The results of this laboratory work must be considered as preliminary as the conclusions have not been thoroughly tested in the field. Clearly it would be prudent to keep particle velocities well below 200 mm/s; indeed for most sites this value may be reduced by a factor of 10 without inhibiting the most efficient progress of the works.

It is intended to check these laboratory tests by site measurements and the next section of this paper describes observations on a newly cast mass concrete wall subject to very high vibration levels. If the laboratory tests are confirmed by a variety of site observations then some relaxation in current restrictions would seem desirable.

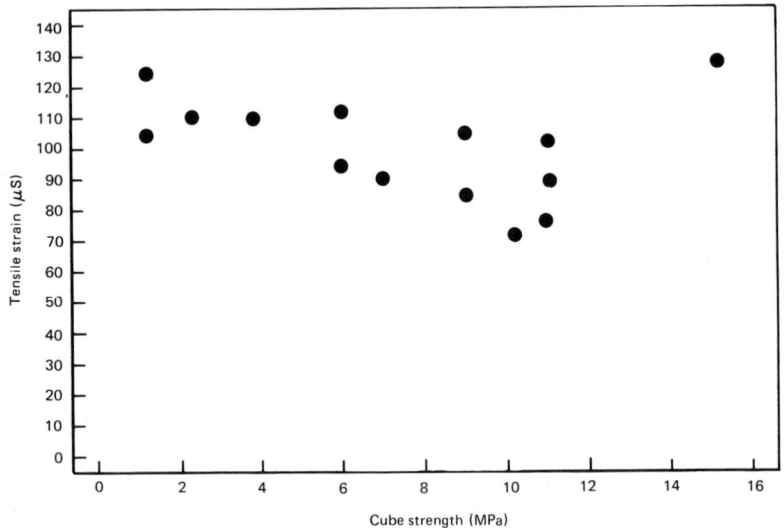

Fig. 13 Strain in concrete prism at failure

PRE-SPLIT BLASTING

In the UK pre-splitting techniques are being increasingly specified for rock cuttings and tunnel portal areas. Carefully designed presplits will provide a relatively undamaged rock face which requires a minimum of maintenance, a factor of considerable importance in the design of major road schemes.

It has been considered that pre-splitting techniques also give rise to high levels of ground vibration (Oriard, 1979) and this is likely to be true where the pre-split is created by a row of holes tightly packed with explosive. In this case there is minimal fragmentation until the subsequent bulk blast removes the burden beyond the pre-split plane. However, a method of pre-split blasting has been developed (Matheson, 1983) which uses air decoupled charges in the plane of the finished rock cut. This method can produce a particularly clean face and has been used on major road schemes in the UK.

A further advantage is that the decoupled charges cause considerably less vibration than similar charges tightly packed. Figure 14 summarises the results of a series of trials carried out by TRRL to determine the effect of various degrees of decoupling. (The decoupling ratio is expressed as the ratio of the drillhole to explosive diameter.) These trials were carried out in strong schists at the site of a new road scheme which required extensive cuttings in rock. Blasting gelatine in plastic sleeves was centred in the

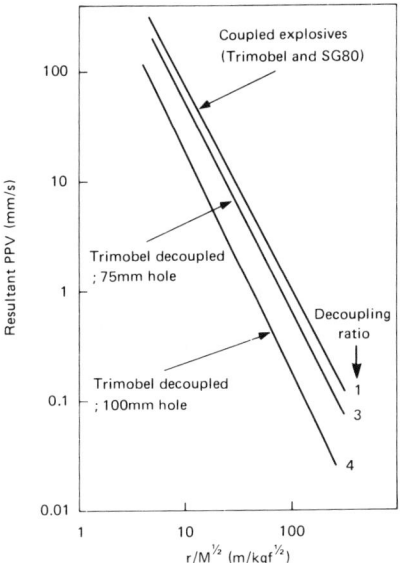

Fig. 14 Peak particle velocity as a function of scaled distance with varying decoupling ratio

drillholes using cruciform spacers with the holes tightly stemmed in the top one metre. The annular space around the explosive substantially reduced the pressure applied to the drillhole and created a mismatch in the effective rock/explosive impedances. These factors significantly reduced the vibration caused by a given charge weight.

The rock above the western portal area of the Penmaenbach tunnel in North Wales was also excavated using similar pre-split techniques. Before the tunnel excavations could begin portal preparation works required the firing of a large pre-splitting charge to yield a near vertical rock face which was damaged as little as possible by the blasting. This would assist with safety aspects by minimising falling rock both during and subsequent to the works and reduce maintenance costs during tunnel operation. Figure 15 shows the tunnel portal with the pre-split face clearly visible. The creation of a pre-split plane some 20 metres wide by up to 14.4 m metres deep required the instantaneous firing of approximately 90 kg of explosives. The main array of pre-splitting drillholes were 90 mm in diameter at 0.9 m centres and were charged with 250 gms/m of 80 per cent strength blasting gelatine fired by blasting cord at 40 g/m. The individual holes were coupled by a continuous surface run of detonating cord which ensured the near simultaneous firing required to create a

Fig.15 The Penmaenbach tunnel portal

successful pre-split plane. The overall charge weight for the presplit plane was approximately 0.32 kg/m^2 of rock face. Transducer arrays were fixed on rock exposures and to a deflector wall which had been recently cast just above the portal. At its closest point the wall was less than 10 m from the pre-split face and very large particle velocities were to be expected.

Peak particle velocities of up to 375 mm/s were sustained by the deflector wall with no visible signs of vibration induced damage. Subsequent ultrasonic pulse velocity measurements on the wall indicated that the concrete strength had not been adversely affected by the blasting. The spectral distribution of vibrational energy is given in figure 16 for the motions of the deflector wall and of a surface rock exposure about 60 m from the blast. Note that for this large and spatially diffuse charge configuration the spectra are dominated by low order harmonics and almost all the energy occurs at frequencies of less than 60 Hz.

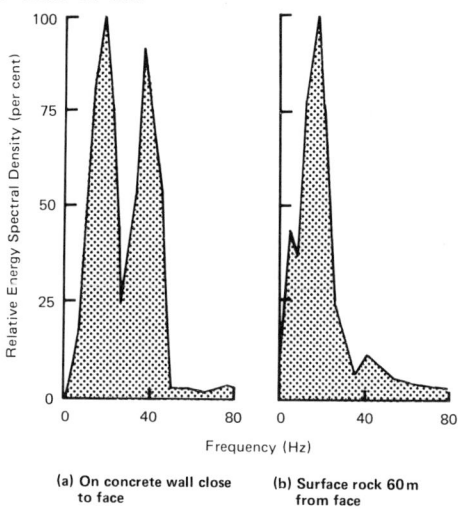

(a) On concrete wall close to face (b) Surface rock 60 m from face

Fig. 16 Energy spectra from presplit explosion

VIBRATION AND SOIL REINFORCEMENTS

In the UK reinforced soil retaining walls are increasingly being considered as cost effective options for use in the construction of major road retaining structures. Whilst it is considered that the vibration caused by the operation of the road is most unlikely to have any serious effects on a well designed soil reinforced structure consideration must be given to the effect of the vibratory compaction plant used during construction. Vibrations can generate substantial dynamic stresses within fill materials. Fluctuations (compression and rarefaction) are thereby imposed on the vertical stress acting between soil and reinforcements, and also on the lateral thrust on the back of the wall.

A reduction in the vertical effective stress produces a corresponding reduction in the resistance to pull-out of the reinforcements, which may result in slippage of the reinforcements and deflection of the face of the wall. Accumulation of such movements could lead to a serviceability failure of the structure.

Research to investigate how compactor induced vibration affects the performance of soil reinforcements in both well and uniformly graded soils has been carried out at TRRL (Brady et al, 1990). The tests measured the reduction in pull-out resistance of galvanised mild steel reinforcements subject to various static and dynamic loading conditions. The tests were carried out in a large test pit at the Laboratory and were designed to cover a range of field conditions. For most tests a constant rate of displacement of 12 mm/hour was maintained until the peak pull-out load was reached. The rate of displacement was then increased to 25 mm/hour and vibration applied to and the surface of the backfill by a large plate compactor. The compactor was moved at a slow walking speed along a line parallel to, and at various distances vertically above, the reinforcement under test. The vibration caused an immediate reduction in the resistance to pullout and was continued until the lower resistance load had reasonably stabilised. After vibration had ceased, then with increasing displacement, the resistance increased to a new reasonably constant level.

The sequence was repeated up to three times on any one strip and for strips at a range of depths. During the tests both the pull-out force and the displacement of the reinforcements were automatically recorded and processed by computer. Readings from stress cells and geophones embedded at various locations and depths in the fill were also recorded so that reductions in pull-out loads could be correlated directly with the variation in static and dynamic stress across the reinforcing elements. Fig 17 shows a typical relationship between the pull-out force and displacement for reinforcement subjected to vibration.

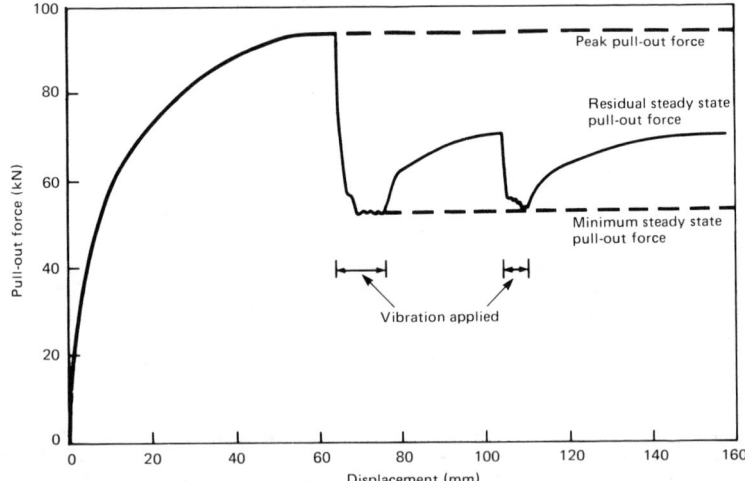

Fig. 17 Typical relation between pull-out force and displacement

The tests showed that the pull-out resistance of planar reinforcements was reduced by dynamic stresses generated by vibration from compaction plant. The reduction in pull-out resistance correlated well with the reduction in normal stress caused by the stress waves generated by the compactor. The relative reduction in pull-out resistance was similar for tests performed in well graded and uniformly graded sands. These trials have provided useful information which allowed the effects of vibration to be specifically accounted for during the design of reinforced soil structures. Further there do not appear to be any over-riding technical reasons why uniformly graded soils should not be considered for use as backfills for these structures.

CONCLUDING REMARKS

Construction techniques which generate significant vibration are likely to be required for the foreseeable future. Environmental factors have become more restrictive and review of levels of vibration considered acceptable indicates a general lowering of permissible limits. This trend is not supported by new field evidence of damage at lower particle velocities and indicates a change in social climate rather than a change in engineering values. Nevertheless it must be expected that more rigorous vibration controls are likely and preconstruction investigations may be required particularly when explosives are specified.

Some Clients and Engineers are attracted by the idea of imposing unjustifiably low limits of vibration on their Contractors. These limits are a form of regressive conservatism which is to be avoided. A common example is the use of the "method of halves", whereby a specifying engineer halves the limit set by his predecessor on a similar job. These methods inevitably impose unreasonable restraints on the Contractor, thereby increasing costs to the Client. Although safe limits must be chosen, overconservative limits and unnecessary restrictions will be a charge on the community as a whole. Therefore, it is vital that authorities responsible for setting vibration limits do so on an informed rather than on an arbitrary basis.

The effects of construction induced vibrations must be considered not only with respect to nearby structures and people but also with regard to the works themselves. For example research reported in this paper indicates that vibration effects on soil reinforced structures may influence design and blasting can cause rock slope failures particularly in the vicinity of tunnel portals. On the other hand laboratory observations suggest that freshly poured concrete may be rather less vulnerable than previously supposed. Vibration levels may be reduced during pre-split blasting without decreasing effectiveness by decoupling the explosives in oversize drill holes.

The correct choice of instrumentation is crucial to the success of vibration monitoring at construction sites. Each link must be considered individually in terms of amplitude capacity and frequency response.

Spectral analysis of vibrations from construction sites in the UK indicate that higher frequencies are present than commonly reported in case histories from elsewhere. At locations close to the explosive source the frequency of the motions may be higher than can be measured reliably with many automatic vibration measuring systems which were often designed to monitor quarry blasting at relatively large ranges. This will be particularly true when small charges are fired in strong rocks. In these circumstances it will be desirable to use high speed data logging devices with sampling rates of up to 10 kHz per channel. At present this type of equipment is operated only by specialist groups and the industry awaits a commercially available seismograph which is appropriate for routine use on construction sites where high frequencies are significant.

Although most structures are relatively unresponsive at high frequencies it is irrational to set limits in terms of particle velocity and then use an instrument incapable of measuring the significant particle velocities present. Where vibrational accelerations approach or exceed "g" no quantity of sandbags will keep the transducers in good contact with the ground. In these circumstances the geophone or accelerometer packages must be rigidly fixed.

When presenting blast vibration data the choice of scaling method should be carefully considered. Blasting conditions can result in scaling relationships which are very different from those implicit in the traditional 'square' or 'cube root' methods. In these circumstances the benefits of the 'site specific' method can be considerable in rationalising the data or providing better predictive equations. The calculation of specific and independent exponents (α and β) describing the influence of range and charge weight can provide information which significantly effects the design and control of the works.

ACKNOWLEDGEMENTS

The work described in this paper forms part of the programme of the Transport and Road Research Laboratory and is published by permission of the Chief Executive. The author gratefully acknowledges the major contributions made by his colleagues at TRRL.
Crown Copyright. The views expressed in this Paper are not necessarily those of the Department of Transport. Extracts from the text may be reproduced, except for commercial purposes, provided the source is acknowledged.

REFERENCES

Ambraseys, N.N. and Hendron, A.J.Jr. Dynamic behaviour of rock masses. Rock Mechanics in Engineering practice. Stagg and Zienkiwicz eds. Wiley, London, 1968, pp 203-236.

Brady, K.C., Watts, G.R., and New, B.M. The effect of vibration on the pull-out resistance of reinforcements in soil. Transport and Road Research Laboratory Research Report 261, TRRL, Crowthorne, UK. 1990.

Crabb, G.I., Hiller, D.M., New, B.M. Ground vibrations caused by the construction and operation of transport systems. FMGM-91, 3rd Int. Symp. on Field Instr. in Geomechanics, Oslo, Balkema, Sept, 1991.

Dowding, C.H. Blast vibration monitoring and control. Prentice-Hall, NJ, USA, 1985.

Matheson, G. Pre-split blasting for highway rock excavation. Transport and Road Research Laboratory Report 1094, TRRL, Crowthorne, UK. 1983.

New, B.M. Trial and construction induced blasting at the Penmaenbach tunnel. Transport and Road Research Laboratory Research Report 181, TRRL, Crowthorne, UK. 1989.

New, B.M. Ground vibration caused by construction works. Tunnelling and Underground Space Technology. Vol 5, No 3, 1990a, pp 179-190.

New, B.M. Ground vibration associated with tunnel construction. Proc. Tunnel Construction 90, Inst. of Min. and Metal. London. 1990b, pp 67-76.

New, B.M. and O'Reilly, M.P. Tunnelling induced ground movements; Predicting their magnitude and effects. Invited review paper to 4th Int. Conf. on Ground Movements and Structures. Cardiff. Institution of Civil Engineers, London. July, 1991.

Newmark, N.M. Problems in wave propagation in soil and rock. Proc. Inst. Soc. Wave Propagation and Dynamic Properties of Earth Materials. Univ. New Mexico Press, Albuquerque. 1968.

Oriard, L.J. Modern blasting in an urban setting. Atlanta Research Chamber Applied Research Monographs. UMTA-GA-06-0007-79-1. June, 1979.

Grouting Techniques For
Excavation Support

Joseph P. Welsh[1], F. ASCE

Abstract

In the design of excavation support for foundations, underground structures and utilities, many construction options are available. These include soldier piles and lagging, sheet piles with tiebacks, structural slurry diaphragm walls, ground freezing and grouting. As new methods of grouting and improvement in grouting materials, equipment, and monitoring systems have been realized, grouting has played an increasingly more important role in this area. This paper will discuss the four main grouting types used for excavation support - slurry, chemical, compaction and jet. The definition, history, design and construction considerations, advantages, limitations, references, relative cost and case histories for each specific types of grouting will be given.

Introduction

Many publications, technical articles and Specialty Conferences (Lambe & Hansen, 1990) have detailed the design, construction and performance of various excavation and earth support systems including grouting. However, the recent advances in grouting technology and their subsequent expanded uses make it apropos for a comparison of slurry, chemical, compaction and jet grouting for excavating support and underpinning. Each type of grouting has a different purpose, uses different materials, different pumping pressures and equipment, and produces a different end result.

Slurry and chemical permeation grouting involve filling voids between the soil particles with either cement or chemical binders. Compaction grouting involves

---
1- Vice President, Hayward Baker, A Keller Company, 1875 Mayfield Road, Odenton, Maryland 21113

densification of loose soils by the injection of very stiff grout. Jet grouting involves pressure fluid jet excavation of soils and replacement with a mixture of the excavated soil and cement grout. Essentially, all types of ground conditions can be controlled utilizing one or more of these grouting techniques with grouting performed in an active mode, modifying the soils being disturbed or in a passive mode, modifying the soils supporting important structures. Each method affects the strength, cohesion, permeability and stiffness properties of different soil types to varying degrees. Figure 1 shows the approximate range of soil types treatable by each method.

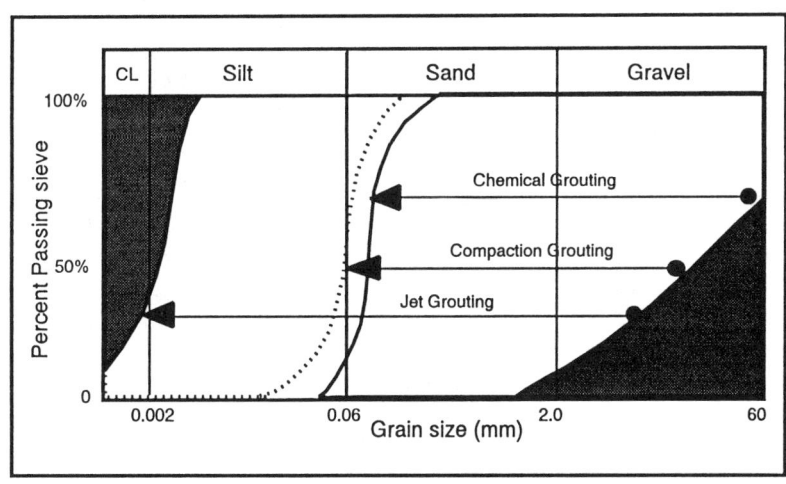

Figure 1 - Range of Soil Types Treatable by Chemical, Compaction & Jet Grouting

Settlement problems anticipated or caused by excavation or tunneling operations have been solved through the use of slurry, chemical, jet and compaction grouting methods, with application either prior to excavation or tunneling to modify the ground to control settlements and/or when difficult ground is unexpectedly encountered. In the design stage, grouting may be specified to strengthen loose or weak soil and prevent crippling cave-ins and ground losses where excavations or tunnels are close to important structures or critical utilities. It may also be specified where adjacent or surrounding structures are supported by shallow foundations, in order to reduce subsidence due to disturbance of loose, sensitive or weak soils. In some cases, grouting may be specified to reduce the ground subsidence effects of dewatering or to prevent the loss of fines from the soil. During actual construction, soil grouting can usually solve problems associated with fine uniform sandy soils that have a tendency to run in a dry state and also weak sandy soils that are unstable in a saturated state (Gularte, 1988).

Some soils can be treated by one type of grouting and not another, but all prospective grouting projects require a detailed soil and soil-structure interaction analysis. All forms of grouting require a two step construction operation; the installation of the grout injection pipe, and the injection of the grout material. Grouting is one of the many techniques that fall under the general category of ground modification or soil improvement. If the preliminary design indicates the possibility of modifying the ground, then a more detailed than normal site investigation should be programmed. Adequate borings should be taken so a continuous profile can be drawn across the site with a high degree of confidence. Continuous standard penetration tests with hydrometer analysis of the fine portion of the sample should be performed. The following will discuss the four types of grouting, their definition, history, design and construction considerations, advantages, limitations, delivery systems, references, case histories, and areas needing additional research.

PERMEATION - SLURRY GROUTING

Definition - Slurry grouting is the intrusion, under pressure, of flowable particulate grouts into open cracks, voids and fractures.

History - Slurry grouting is the oldest of the major grouting techniques, with a history dating from the development of grout pumping equipment in the 1870's. Weaver (1991) reports that the earliest use of slurry grouting in the United States was the injection of fissured rock beneath a dam in New York State in 1893. Its primary use has been to reduce the permeability of rock beneath dams and underground structures, and these uses have contributed to slurry grouting being the largest volumetric user of grouts in North America.

Despite slurry grouting's long history, the technique has generated considerable and continuing controversy in the engineering community. The relatively slow evolution of the system, without benefit of actual research and development, has resulted in many early "rule of thumb" approaches which are still currently accepted as rules of practice, but nowadays, are being increasingly challenged.

One major contention is the proper water-cement ratio to use with cementations grouts. High water contents allow more grout to be injected, but these grouts then have the potential for shrinkage, resulting in poor quality of grouts in-place. Two papers in the April 1985 ASCE publication, "Issues in Dam Grouting," (Baker, 1985), one by Houlsby, "Cement Grouting: Water Minimizing Practices;" and the other by Deere and Lombardo, "Grouting Slurries, Thick or Thin?," spell out this controversy in depth. Two recent books are must reading for any engineer involved with slurry grouting, "Construction and Design of Cement Grouting" by Houlsby (1990) and "Dam Foundation Grouting" by Weaver, (1991).

Design and Construction Considerations - The process of permeation grouting involves a number of steps for success. Defining the problem, the soil or rock types present, the modified soil strength desired, the array of grout pipes to cover the design zone and the injection procedure. The quality assurance, quality control and costs associated with the grouting process also need to be addressed for each given design situation.

When injecting into coarse soils, a grout composed of normal cement, bentonite and filler is typically used. A filler of flyash or fine sand is often used to reduce material costs. Bentonite helps maintain the cement in suspension during grouting and can also be used to control the final strength of the grout.

The main consideration for slurry grouting in sealing cracks and fissures in rock or injecting soils for either water control or structural improvement purposes is the grain size of the particulate grout compared to the width of the rock fracture or the grain size of the soil to be grouted. Mitchell (1981) presented groutability ratios for soils and rocks as follows:

For soils: $N = \dfrac{(D_{15}) \text{ soil}}{(D_{85}) \text{ grout}}$

$N > 24$: grouting consistently possible
$N < 11$: grouting not possible

$N_c = \dfrac{(D_{10}) \text{ soil}}{(D_{95}) \text{ grout}}$

$N_c > 11$: grouting consistently possible
$N_c < 6$: grouting not possible

For Rock: $N_R = \dfrac{\text{fissure width}}{D_{95}) \text{ grout}}$

$N_R > 5$: grouting consistently possible
$N_R < 2$: grouting not possible

Where "D" equals the soil or grout diameter, and its subscript equals the percent finer.

We cannot change the soil or rock characteristics but we can control the fineness of the particulate grout. The introduction of fine grind cement grouts allows for better grout penetration (Clarke, 1984). Figure 2 gives gradation curves for conventional Portland Cement and some fine grind cements.

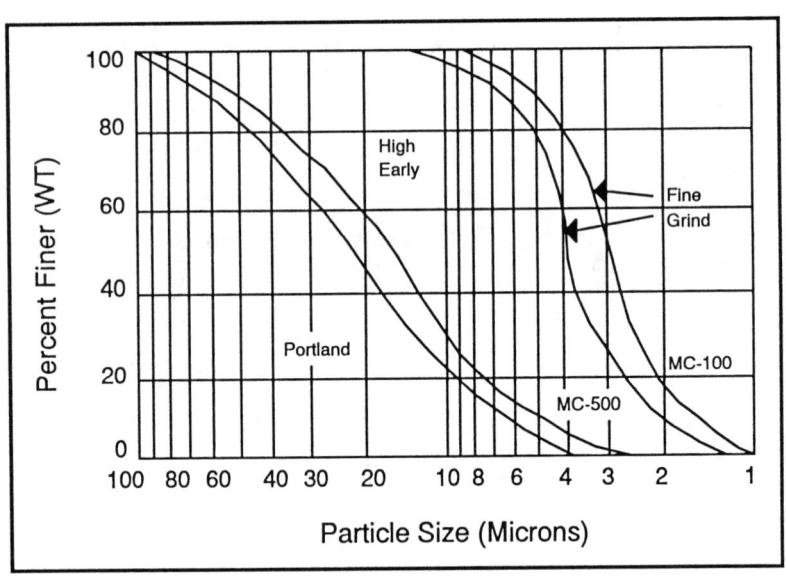

Figure 2 - Gradation Curves for Conventional Portland and Fine Grind Cements

Another advance in slurry grouting is the ability to monitor the flow of grout through the fractures in the rock by the utilization of acoustic emission technique (Koerner et.al., 1984); however, some additional research is necessary before this technique becomes a standard of practice. Albreitton (1982) states the cement grouting practices of the United States Army Corps of Engineers and suggests a starting mix design of water cement ratio of 6:1. However, Houlsby (1985), Deere and Lombardi (1985) and others suggest lower water cement ratios throughout the grouting operation. "Design of Cement-Based Grouts," (Littlejohn, 1982) is an excellent paper on the proper design of cement-based grouts covering constituent materials, bleeding, flow properties, setting, shrinkage, thermal properties, grout strength and durability.

Grouting Equipment - Most experts agree that high speed colloidal grout mixers are superior to the standard slow speed mechanical mixers. Colloidal grout mixers produce grout of greater uniformity with better penetrability and pumpability. Cement clusters are separated and the individual particles are often broken and rounded to a significant degree, and colloidal mixers make it possible to grout tighter fractures than standard mixers. Colloidal mixers are also required for mixing and hydrating bentonite.

There are three basic types of grout pumps: the piston pump, the diaphragm pump and the helical screw pump. Many government agencies prefer the progressive cavity or helical screw pump because they can maintain a constant pressure without surging and will work at operating pressures up to 700 psi (4.8 MPa). Gourlay and Carson (1982) give a good description of slurry grouting plant and equipment. As the initial set does not occur in normal slurry grouts until after three hours, an instrumentation program has to be established for injection of these grouts to preclude surface heave, objectionable fracturing and loss of grout. Careful monitoring of flow versus grouting pressure has to be maintained.

Although the most economical of the grouting techniques, rarely is a soil formation uncovered that is both permeable and uniform enough to allow treatment for excavation support by slurry grouting.

PERMEATION - CHEMICAL GROUTING

Definition - Chemical Grout - "Any grouting material

characterized by being a pure solution; no particles in suspension" (ASCE Preliminary Glossary, 1980).
History - In his 1990 second edition "Chemical Grouting," Karol traced the history of chemical grouting and indicated that Joosten is credited with the earliest demonstration of the reliability of the chemical grouting process with his patented Joosten Process in 1925. This basically involved the injection of concentrated sodium silicate into one hole and a strong calcium chloride solution under high pressure into an adjacent hole. In the early 1950's, AM-9 was introduced into the U.S. marketplace by American Cyanamid Company. American Cyanamid's Research Center published over 1,000 pages of technical reports relating to chemical grouting based on their extensive research. AM-9 was a mixture of organic monomers polarized insitu, and the control of gel time could be designed from seconds to hours. In the early 1980's, AM-9 was removed form the market due to potential neurotoxic problems with the grout and it was subsequently replaced by a low toxicity AC-400 grout. In the early 1960's, a single shot sodium silicate-based grout, trademarked Siroc, was introduced into the market by Diamond Alkali Company and extensive research was also published on this product.

The Federal Highway Administration, anticipating a large utilization of chemical grouting for various proposed subway systems, commissioned extensive research and reports including: a two volume report in 1976 by Hendron and Lenehan entitled, "Grouting in Soils," Volume I being a State-of-the-Art Report, and Volume 2 being a Design and Operation Manual; in 1977, Tallard and Carson's two volume report, "Chemical Grouting for Soils,"covered available chemical grout materials and an engineering evaluation of these materials; and in 1982, a four volume report by Baker et.al. was published on the "Design and Control of Chemical Grouting." The ten Rapid Excavation and Tunneling Conferences have included numerous papers on various projects where grouting has assisted construction including a paper by Clough et.al. (1979) on ground control for soft ground tunneling by chemical stabilization. The 5th RETC conference included a paper (Puza et al, 1981) discussing the use of cement, chemical and compaction grouting for the mixed face portion of the Red Line Extension through Somerville, Massachusetts.

"Tunneling Performance of Chemically Grouted Alluvium and Fill, Los Angeles Metro Rail, Contract A-130" (Gularte, et al 1991) describes the largest use of chemical grouting in the United States with the injection of two million gallons (7.57m liters) of chemical grout. The program was more than justified by the minimal movement of the 10 lanes of the Hollywood Freeway during the tunneling of the twin 21 foot (6.4 meter) diameter tubes beneath the freeway. However, prior to the final concrete lining, a fire broke out in the YL tube consuming all the timber lagging and damaging the steel rings, leaving the chemical grouted soil as the only support. Still, no differential movement occurred and the decision to chemical grout was further justified, particularly when a non-critical ungrouted section of the tunnel daylighted due to lack of support.

In the late 1970's, polyurethanes were introduced into the grouting market and, although their costs are relatively high, these offer excellent waterproofing capabilities. Welsh's 1984 paper lists and discusses commonly used grouts for control of water infiltration in underground transportation structures. Over 85 percent of chemical grouting is performed for strength producing purposes, in effect adding cohesion to sands to increase stability and prohibit running during underground construction operations. Sodium silicate grouts, with various catalysts and reactants, are used for this purpose. Welsh's 1983 paper describes chemical grouting of a "bathtub" inside an existing commercial building to allow construction of a machinery pit adjacent to heavily loaded retaining walls and building footings. The sodium silicate grout effectively underpinned the footings and permitted excavation without dewatering of the granular formation despite a high ground water table. The other main use of chemical grouting is for water control purposes either in soil, rock or underground structures, frequently after construction when various factors have allowed water to seep into these facilities.

Design and Construction Consideration - Figure 3 presents grain size curves and depicts the normal area of soil that can be injected with chemical grouts. Coarser soils are normally injected with cement grouts, while sands with more than 15 percent passing the 200 sieve are too clogged with this fine grained materials to allow economical permeation with chemical grout. In 1982, Baker published an excellent paper on "Planning and Performing Structural Chemical Grouting," presenting the chemical groutability of the soil by its permeability. He indicated that soils having a permeability in the range of 10^{-1} cm per second to 10^{-2} cm per second are easily groutable; soils showing permeabilities in the range of 10^{-3} cm per second to 10^{-4} cm per second are moderately groutable. When permeability is from 10^{-4} cm per second to 10^{-5} cm per second, the soil is usually only marginally groutable and may be ungroutable from a practical point of view. Soils with permeability lower than 10^{-5} cm per second are considered ungroutable. All grouting operations consist of installation of grout pipes and then the controlled injection of the grout

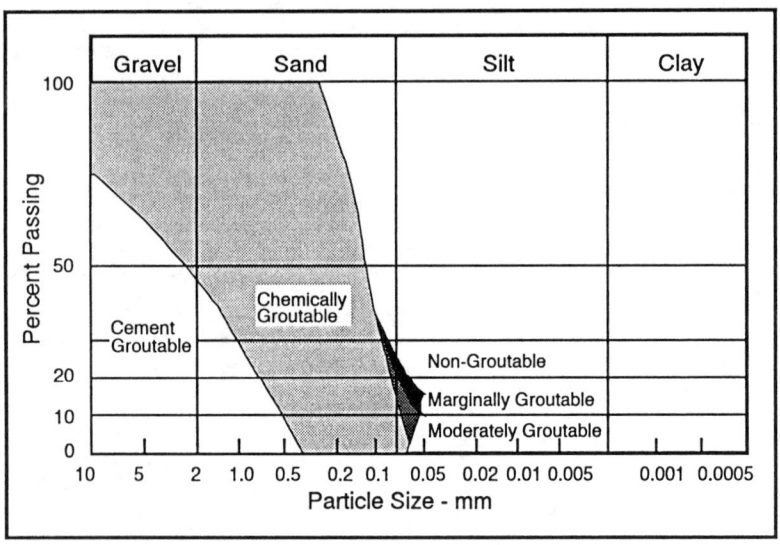

Figure 3 - Grain-Size Ranges for Chemically Groutable Soils

through these pipes. The sleeve port pipe, as shown in Figure 4, provides the most versatile grout injection system, particularly for chemical grout. By locating the internal grout packer at a particular sleeve port, grout can be injected at any specified depth and regrouting can be performed at that depth in any number of stages. Distinct advantages occur with the use of the sleeve port pipe: 1) multiple grout injection point with depth increases the likelihood of saturating the desired area; 2) the elastic sleeve at each port prevents the grout from returning into the grout pipe and gelling up after grouting, thus allowing the port to be reused; 3) the sleeve port pipe is completely sealed in the borehole, reducing grout leakage away from the desired zone along either the pipe or bore hole interface. The installation of the sleeve port grout pipe involves drilling the hole, installing the sleeve port grout pipe and then placing the sealing grout to surround the entire pipe and fill the annulus. The grout used to fill the annulus is usually made up of cement, bentonite, and flyash. The grout should be thick enough to prevent infiltration into the soil, and ideally, it should be low strength and brittle. The pipe should be positioned below ground surface and a protective cap provided where pedestrian, vehicle traffic and/or vandalism is possible.

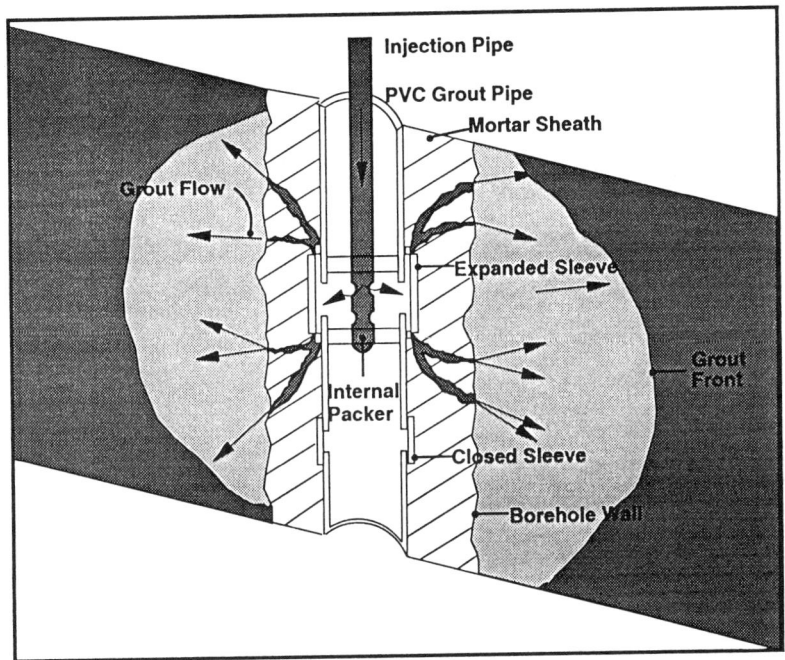

Figure 4 - Sleeve Port Pipe

If a sand stratum is predetermined to be groutable, an estimated cost for making the sand into a sandstone by chemical grouting can be determined by taking the volume of sand to be treated, assuming a 30 percent grout take and figuring two to five dollars per gallon (3.8 l) for the chemical grout to be injected into sleeve port pipes. The cost of installation of the sleeve port pipes would normally be in excess of fifteen dollars per linear ft (0.3 m) depending upon drilling conditions, number of pipes to be installed, labor costs, etc. As previously stated, sleeve port grout pipes provide better control of the chemical grouting injection and allow primary, secondary and tertiary grouting of a zone. Normally, 70 percent of the grout is calculated to be injected into the primary holes and the remainder into the secondary and tertiary grout holes. By properly maintaining and reviewing the grout injection records, the secondary and tertiary grouting can serve as measures of the success of the grouting operation.

In general, the lower the soil permeability, the slower the injection rate and the greater the grouting pressure is required. The higher the pressure, the greater likelihood of accidental hydraulic fracturing the soil during grouting. Every soil has an optimum rate of injection for a given pumping pressure. If grout is injected at too high a pressure for a given depth, fracturing will occur. If pumped too slowly, an uneconomical project will result. For chemical grouting, the injection rate is often found to be between 0.5 and 6 GPM (0.03 and 0.38 l/s). Some controlled soil fracture toward the end of the injection process is usually considered necessary to achieve complete grout penetration. In areas where drill accessibility is difficult or where a heavy concentration of utilities exist, installation of grout pipes horizontally or at angles may be necessary to reach the problem soil.

COMPACTION GROUTING

Definition - In 1980, the ASCE's Committee on Grouting, in its preliminary glossary, gave the following definition: "Compaction Grout - Grout injected with less than one inch (25 mm) slump. Normally, a soil cement with sufficient silt sizes to provide a plasticity together with sufficient sand sizes to develop internal friction. A grout generally does not enter soil pores but remains in a homogeneous mass that gives controlled displacement to compact loose soils, gives controlled displacement for lifting of structures or both."

History - Most of the ground modification systems used in the United States have been first introduced overseas and then later to the United States market; this is mainly due to the separation of responsibility between design and construction in the United States. Compaction grouting is the only ground modification technique developed in the United States which is now being exported overseas.

Warner's 1982 paper, "Compaction Grouting - The First Thirty Years," describes compaction grouting and its uses mainly as a technique for remedying settlement of slabs, buildings, etc. In 1983, Baker et.al. described the Bolton Hill Tunnel, part of the Baltimore Subway system, where compaction grouting was first used in lieu of underpinning to protect 39 structures. Subsequent compaction grouting has been used in this application for many of the soft ground tunnels in the United States including the Seattle Light Rail Transit system.

The underground construction industry in the United States tends towards performance type specifications, thus allowing for easier innovation in this area than in the building construction industry.

Design - The key to success in compaction grouting is the proper mix design. One of the advantages that compaction grouting has over slurry grouting is that the low slump grout being injected gives the designer and contractor better control over the grouting operation. In idealistic situations, the degree of improvement in densification required at a particular stratum can be predetermined and the correct volume of compaction grout injected to meet the design requirements. Salley et.al., (1987) describes a compaction grouting test program at the Pinopolis West Dam in South Carolina. This case history describes how an existing loose, liquefiable sand strata was improved by a predetermined amount by compaction grouting. Like any grouting operation, the compaction grouting operation can be broken into two phases, the installation of the grout pipes and the injection of the grout. Due to the low slump grout injected and the pipe friction generated, a minimum of a two inch (50.8 mm) diameter pipe is required with, more realistically, three inch (76.2 mm) diameter pipe used. Although there has been some advocation of grouting in stages from the top down for redensification of loose soil beneath structures and other remedial work, from an economic standpoint the majority of the work is performed by placing the pipes to the maximum depth and then injecting the low slump grout under a controlled pressure.

A pre-grouting system is established to monitor heave, grout pressure and volume, with limits on each project depending upon specific site requirements. The spacing of grout pipes will depend upon the end results desired, but normally 5 ft (1.5 m) on center is the minimum spacing and, on some densification projects, primary grout pipe spacings have been 25 ft (7.6 m) on center with secondary and, if needed, tertiary pipes on split spacings. Pumping a low slump grout will cause a large friction loss through a grout pipe, and in order to ascertain actual pressure at the bottom of the hole, the mix utilized can be monitored with various hose and pipe length and pressure gauges established at the pump header and tip of the grout pipe prior to actually placing the pipes into the ground.

As indicated by the ASCE 1978 publication, "Slabjacking - A State of the Art Report," compaction grouting can be effectively used to jack slabs back to original elevation. Also, buildings have been raised, and loose soils densified beneath settling buildings and piles.

In compaction grouting adjacent to underground structures, careful monitoring has to be performed as tunnel lagging has been snapped, and underground boring machines have assumed an "egg shape" by the high pressures of compaction grouting operation.

More and more soft ground tunneling projects are protecting adjacent structures by compaction grouting in lieu of the more expensive and often more settlement-prone pit method of underpinning. By specifying compaction grouting to be performed, the grout pipes can be pre- placed and the structures to be protected can be carefully monitored so that if any objectionable movement takes place, compaction grouting can be used to immediately rectify the loss of soil and prevent further movement.

Like all grouting programs, the cost of compaction grouting can be broken down into 3 basic element: mobilization / demobilization, grout pipe placement, and injection of compaction grout.

> Mobilization / Demobilization: The cost can start under ten thousand dollars and reach fifty thousand dollars when multi-drill rigs and large batch plants are required.
>
> Grout Pipe Placement Cost: Three inch (76.2mm) diameter grout pipes are normally installed on a primary / secondary grid pattern at costs between fifteen and sixty dollars per linear meter.
>
> Compaction Grout Injection Cost: The volume of grout injected normally ranges between 5 and 20 percent of the total mass volume being treated and the injected cost starts at two hundred dollars per cubic meter of grout injected.

JET GROUTING

Definition - Jet grouting is a procedure for the insitu construction of solidified ground of a predetermined shape, size and depth to design characteristics, e.g., strength, permeability and/or flexibility. The insitu product is called Soilcrete.

History - Jet grouting was initially developed in Japan in the early 1970's, introduced into Europe in the late 1970's and the United States in the early 1980's. It has been used for underpinning and/or excavation support of sensitive structures, groundwater cut-off control and for tunneling applications. Jet grouting has been performed horizontally to support the tunnel roof, to ease boring operations and to protect against subsidence due to soil loss.

Jet grouting's effectiveness across the widest range of soil types, including silts and some clays, has established it as a valuable addition to the range of grouting techniques.

Burke et.al. (1989) described the first documentation of jet grouting in North America used both for excavation support and underpinning.

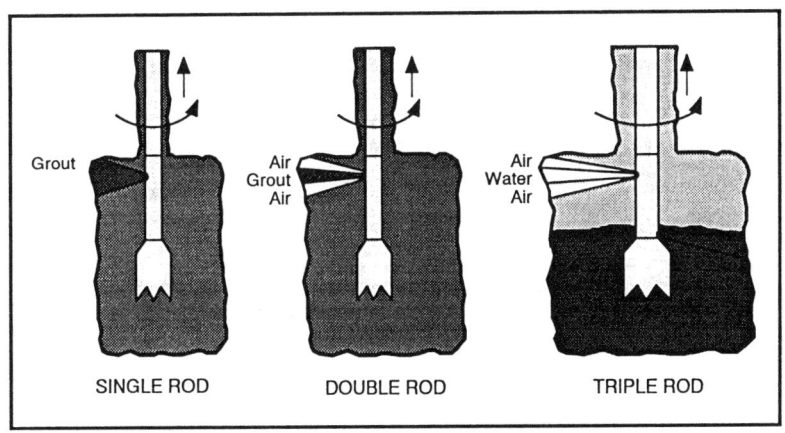

Figure 5 - Jet Grouting Systems

Three basic systems have been developed - the single, double, and triple rod techniques (Figure 5). The single rod method uses high pressure cementations slurry material to both cut the soil and form a soil cement matrix. The double rod system horizontally jets a cementations slurry material protected and assisted by an air sheath, producing larger diameters. Technically the most advanced, the triple rod system, uses a combination of high pressure water shielded in a cone of air to cut and lift the soil to the surface while simultaneously filling the created cavity with a pre-engineered cement slurry. Depending on the lift and injection parameters selected, the system can be designed to mix the soil with a grout or remove the soil and replace it with grout.

The triple rod system allows for the best control of the unconfined compressive strength, since the majority of the fine grained soil is removed and replaced with a premixed material, unlike the single and double rod systems which incorporate the soil matrix as part of the cementations material. Therefore, the triple rod system has been used extensively for underpinning and where strength is required. Shafts have been supported by placing interlocking cylinders around the circumference of the proposed shaft excavation. Although the interlocking column configuration is the most frequently used, any jet grouted geometry can be formed to any depth, from any degree of an arc to a continuous panel, with the size of the composite grouted mass dictated by the application required.

Design and Construction Considerations - Like any engineered product, the objectives of the jet grouting operation have to be initially defined and the advantages and limitations of each jet grouting system recognized.

A detailed engineering assessment of soil conditions is paramount to the jet grouting operation. The width or diameter of each constituent panel or column is determined at the design stage by reference to the soils warranting treatment. Operating parameters of air, water and grout flow and pressure, together with monitor rotation and withdrawal speed, are then determined by the geometry of the design column.

In the construction stage, careful maintenance of waste flow to the surface is vital so that pressure build up and heave is avoided. The volume of waste created needs to be addressed. Typically, waste is disposed of off-site or can be utilized as on-site fill. Since only inert compounds are produced by this technique, disposal considerations are volume-related only.

As well as the advantages offered by jet grouting's ability to work well in all soils, the technique also offers high compressive strength and low permeability, no harmful vibrations during installation, and the ability to form any required cross- section. Since only a limited work space is required for the lightweight grouting equipment, restricted access is not a problem.

Due to the sophisticated drilling and computerized grouting equipment, the mobilization cost of jet grouting begins at thirty five thousand dollars. The cost of an element one meter in diameter is normally greater than two hundred and fifty dollars per linear meter. This excludes the handling of the waste created by the soil removal in the jet grouting operation.

Case History - Downtown Bus Transit Tunnel, Seattle, Washington

To prevent settlement of overlying and adjacent structures during construction of the 5,133 ft (1,565 m) long, 21 ft (6.4 m) diameter twin soft ground tunnels, compaction grouting was specified. Although many strata were uncovered during the tunneling operation, these twin tubes were basically driven through over consolidated glacial deposits and glacial outwash. Where the tunnels started, the alignment went directly beneath a 2-story structure and, at the opposite end, a 500 ft (152.4 m) radius went directly beneath a 4-story structure.

To protect the buildings directly over the tunnels, two rows of compaction grout pipes were hand drilled from the basements o the buildings on approximately five ft (1.5 m) centers, with additional holes located in the immediate area of the foundation and in zones where the pillar between the tunnels was narrow. The buildings were supported on spread footings which were located between 12 and 20 ft (3.7 and 6.1 m) above the tunnel crown. In order to avoid problems with the tunneling machine but still redensify the soil, the compaction grouting started at approximately eight ft (2.4m) above the tunnel crown as the shield pass and the temporary lining was erected. Structures adjacent to the tunneling were protected by placement of angled pipes from street level between the tunnel crown and the building's spread foundations, with spacing of these pipes on 7 to 10 ft (2.1 to 3 m) centers. Utilities beneath the streets were protected with compaction grout pipes placed on approximately 10 ft (3 m) centers directly over both alignments.

As tunneling progressed, unanticipated saturated silty sand layers were encountered at several locations, resulting in extensive ground loss at the face of the shield. Quick remedial action was undertaken by pumping replacement material into the loose ground area over the shield through the pre-placed compaction grouting pipes. Later, as the shield passed and the temporary lining was erected, the pipes were used to perform planned compaction grouting.

In one area of the east bound tunnel, ground loss in the vicinity of an adjacent building was too extensive and too rapid to be replaced by compaction grouting. Because the building was supported by a deep foundation, damage was minimal but the incident prompted further investigation of the foundation soils of a second building adjacent to the west bound tunnel where similar ground loss would cause excessive building settlement. In this silty sand, over 70,000 gallons (264,950 l) of a sodium silicate-based chemical grout was injected to stabilize these soils.

Specifications dictated maximum allowable settlement of 0.015 ft (4.5 mm) at the building line along the street and in order to insure that these specifications were met, a compaction grout mix was designed without cementations material so that if later monitoring showed any additional movement, these pipes could be used to redensify and possibly raise the settlement structures.

As the crown of twin tunnels were only 5 ft (1.5m) below the 38 ft (11.6m) high and wide Burlington Northern Railroad Tunnel, it was elected to underpin this structure by jet grouting. The jet grouting limited settlement of the railroad tunnel to less than 0.2 in (0.5cm) during the passage of the shields to construct the twin tunnels beneath the railroad tunnel The details of the project are presented in RETC publications (Robinson, et al, 1991 and Critchfield and Mac Donald, 1989).

SUMMARY

In addition to the case histories previously documented, there are many other projects where grouting has been successfully utilized for excavation support and underpinning and not documented.

The ongoing extension of the Baltimore Subway used jet grouting for water control to compliment the soldier pile and lagging for the Shot Tower Station in the section where utilities precluded the use of a structural diaphragm wall. Chemical grouting protected utilities from settlement when the twin tubes were tunneled under them, and compaction grouting was used to protect buildings from settlement in the zone of influence of the tunnels.

In summary, slurry grouting can cement coarse sands and gravels to prevent unraveling during adjacent excavations; rarely are coarse enough deposits uncovered so that slurry grouting can be used as the prime excavation support. In sands containing less than 15 percent silts or clays, chemical grouting can add sufficient cohesion so that for shallow excavations it can be the prime excavation support and underpinning method. Jet grouting's capability to be effective in most soil types and to produce a high and predictable strength end product directly under existing foundations, makes this system the ideal answer to many excavation support and underpinning projects, particularly when the two uses are combined. Compaction grouting has rarely been used for excavation support, but is an excellent alternative to conventional underpinning. If soils or structures may be disturbed by adjacent construction, compaction grouting can be specified to redensify the soils and relevel the structures, if required.

CONCLUSION

As the latest grouting technologies are utilized and documented in the technical literature, grouting's value as a construction tool is enhanced to the design - construct team. With a detailed soil exploration program, the consultant has the option of specifying a procedure or performance grouting specification. Budget cost can be accurately determined so that the economics of grouting can be compared with alternative systems. Further, if the actual extent of the work can be outlined, the grouting phase can be bid on a lump sum basis with overrun and underrun options. From a design standpoint, the required improvement in the soil can be specified after conventional design principals are applied to the actual loads that will be incurred by the grouting system. Finally, with the proven performance of the jet grouting techniques, grouting has become less dependent on the characteristics of the in-place soils.

REFERENCES

Albreitton, J.A., (1982). "Cement Grouting Practices". U.S. Army Corps of Engineers, Proceedings of the Conference on Grouting in Geotechnical Engineering, W.H. Baker (Editor), Geotechnical Engineering Division ASCE, New Orleans, LA, 264-278.

"ASCE Preliminary Glossary of Terms Relating to Grouting", (1980). Committee on Grouting, Journal Geotechnical Engineering Division, ASCE Proceedings, Vol. 106, No. GT7, New York, NY, 803-815.

Baker, W.H., (1982). "Planning and Performing Structural Chemical Grouting". ASCE Specialty Conference, Grouting in Geotechnical Engineering, New Orleans, LA.

Baker, W.H., Cording, E.J., and MacPherson, H.W., (1983). "Compaction Grouting to Control Ground Movement During Tunneling". Underground Space, Permagon Press Ltd., Vol. 7, 205- 212.

Baker, W.H., Huck, P.J. and Waller, M.J, (1982). "Design and Control of Chemical Grouting, Vol. 1, Construction Control", U.S. D.O.T., Report No. FHWA/RD-82-036; Krizek, R.J. and Baker, W.H., "Materials Description Concepts, Vol. 2", Report No. FHWA/RD-82-037; Baker, W.H., "Engineering Practice, Vol. 3", Report No. FHWA/RD-82-038, Baker, W.H., Executive Summary, Vol 4., Report No. FHWA/RD-82-039, Washington, DC.

Baker, W.H., (1985). "Issues in Dam Grouting". Proceedings of the Geotechnical Engineering Division, ASCE, Denver, CO, 167 pp. U.S. Army WES Information Center and Concrete Laboratory, Vicksburg, MS, 325 pp.

Burke, G.K., Johnsen, L.F., and Heller, R.A., (1989). "Jet routing for Underpinning and Excavation Support". Proceedings of the 1989 Foundation Engineering Congress, F.H. Kulhawy (Editor), ASCE Geotechnical Engineering and Construction Division.

Burke, G.K. and Welsh, J.P. (1991). "Jet Grouting - Uses for Soil Improvement" Proceedings of the Geotechnical Engineering Congress 1991, Geotechnical Special Publication No. 27, ASCE, NY, NY, pp. 334-345.

Clarke, W.J., (1984). "Performance Characteristics of Microfine Cement". ASCE Convention, Atlanta, GA, 14 pp.

Clough, G.W., Baker, W.H., and Mensah-Dwumah, F., (1979). "Ground Control for Soft Ground Tunnels using Chemical

Stabilization - A Case History Review". Proceedings 1979 Rapid Excavation and Tunneling Conference, Vol. 2, Atlanta, GA.

Critchfield, J.W., and MacDonald, J.F., 1989, "Seattle Bus Tunnel Construction," Proceeding, RETC, New Orleans, LA, Vol. 1, pp. 341-359.

Deere, D.U., and Lombardi, G., (1985). "Grout Slurries - Thick or Thin?" Issues in Dam Grouting, W.H. Baker (Editor), Proceedings of the Geotechnical Engineering Division ASCE, Denver, CO, 156- 164 pp.

Gourlay, A.W., and Carson, C.S., (1982). "Grouting Plant and Equipment". Proceedings of the Conference on Grouting in Geotechnical Engineering, W.H. Baker (Editor) Geotechnical Engineering Division ASCE, New Orleans, LA, 121-135.

Gularte, F.B., (1988). "Soil Grouting for Tunneling". Grouting Practices for Shafts, Tunnels and Underground Excavation.

Gularte, F.B., Taylor, G.E., Monsees, J.E., and Whyte, J.P. (1991). "Tunneling Performance of Chemically Granted Alluvium and Fill. Los Angeles Metro Rail, Contract A-130" Proceedings of the Rapid Excavation and Tunneling Conference, Seattle, Washington, June 1991, Society for Mining, Metallurgy and Exploration, Littleton, CO, pp. 291-307.

Hendron, J., and Lenehan, T., (1976). "Grouting in Soils". U.S. D.O.T., Vol. 1 and 2, Report FHWA-RD-76-26 and 27, Washington, DC.

Houlsby, A.C., (1985). "Cement Grouting: Water Minimizing Practices". Issues in Dam Grouting, W.H. Baker (Editor), Proceedings of the Geotechnical Engineering Division ASCE, Denver, CO, 34-75.

Houlsby, A.C. (1990). "Construction and Design of Cement Grouting". John Wiley & Sons, Inc., New York, NY, 442 pgs.

"Innovative Cement Grouting", 1984. J.P. Welsh (Editor), ACI Publication SP-83, Detroit, MI, 175 pp.

Karol, R.H., 1990. "Chemical Grouting" Second Edition, Marcel Dekker, Inc., New York, NY, 465 pp.

Koerner, R.M., Learied, J.D., and Welsh, J.P., (1984). "Uses of Acoustic Emissions as a Non-Destructive Testing

Method to Monitor Grouting". Innovative Cement Grouting, J.P. Welsh (Editor), ACI Publication SP-83, Detroit, MI, 85-102.

Lambe, C. and Hansen, A. (1990), "Design and Performance of Earth Retaining Structure" Proceedings of a Conference sponsored by the Geotechnical Engineering Division of ASCE held at Cornell University, June 18-21, 1990. Geotechnical Special Publication No. 25, 904 pgs.

Littlejohn, G.S., (1982). "Design of Cement Based Grouts."Proceedings of the Conference on Grouting Geotechnical Engineering, W.H. Baker (Editor), Geotechnical Engineering Division ASCE, New Orleans, LA, 35-48.

Mitchell, J.K., (1981). "Soil Improvement - State-of-the-Art Report." Proceedings of the Conference on Soil Mechanics and Foundation Engineering, Stockholm, Sweden, 509-565.

Moller, D.W., Minch, H.L., and Welsh, J.P., (1984). "Ultrafine Cement Pressure Grouting to Control Groundwater in Fractured Granite Rock", Innovative Cement Grouting, J.P. Welsh (Editor), ACI Publication SP-83, 129-151.

Munfakh, G.A. (1991). "Deep Chemical Injection for Protection of an Old Tunnel" Deep Foundation Improvements : Design, Construction, and Testing papers presented at the symposium on Design, Construction, and Testing of Deep Foundation Improvement : Stone Columns and Other Related Techniques, Esug & Bachus editors, ASTM, STP 1089, Philadelphia, PA, pp. 266-278.

Puza, D.E., Borggaard, R.C., Bhore, J.S., Keville, F.M. (1981). "Mixed Face Tunnel Excavation Using Floating Crown Bar and Modified Spilling Method". Proceedings of the 1981 Rapid Excavation and Tunneling Conference, American Institute of Mining, Metallurgical and Petroleum Engineers, Inc. & ASCE, New York, NY, pp. 383-392.

Robinson, R.A., Kucker, M.S., and Parker, H.W., (1991), "Ground Behavior in Glacial Soils for the Seattle Transit Tunnels," RETC Proceedings, June 1991, pp. 93-117.

Salley, J.R., Foreman, B., Henry, J.F., and Baker, W.H., (1987). "Compaction Grout Test Program - West Pinopolis Dam." Soil Improvement - A Ten Year Update, J.P. Welsh (Editor), ASCE Geotechnical Special Publication No. 12, 245-264.

"Slabjacking - State-of-the-Art", (1977). ASCE Committee on Grouting, Journal of the Geotechnical Engineering Division, GT-9, New York, NY, 987-1005.

Tallard, G.R., and Carson, C., (1977). "Chemical Grouts for Soils". U.S. D.O.T., Contract FHWA-FH-11-8826, Vol. I and II, Report FHWA/RD/77/60.

Warner, J., (1982). "Compaction Grouting - The First Thirty Years." ASCE Specialty Conference Grouting in Geotechnical Engineering, New Orleans, LA, 694-707.

Weaver, K.D. (1991). "Dam Foundation Grouting" published by ASCE, New York, NY, 178 pgs.

Welsh, J.P., (1983). "Chemical Grouting Utilized for Underpinning and Water Control". Improvement of Ground, Proceedings of the 8th European Conference on Soil and Foundation Engineering, Helsinki, Finland, 177-180.

Welsh, J.P., (1984). "Control of Water Infiltration by Injection Techniques for Underground Transportation Structures". Tunnel Seepage Control Session, APTA Rapid Transit Conference, Baltimore, MD.

SUBJECT INDEX
Page number refers to first page of paper.

Arbitration, 46

Bids, 46
Blasting, 172, 181, 212
Bracing, 119

Case reports, 65, 119, 240
Chemical grouting, 107, 240
Chicago, 119
Clays, 119
Compaction grouting, 107, 240
Concrete, 212
Conflict, 46
Construction, 212
Construction costs, 1, 172
Construction management, 181
Contamination, 26
Control, 181

Deep foundations, 119
Design, 1, 91, 240
Dewatering, 144

Earth reinforcement, 212
Environmental impact statements, 6
Environmental impacts, 6
Environmental issues, 1, 26, 65
Excavation, 6, 26, 65, 119, 144, 240

Frequency, 181

Great Britain, 65
Ground motion, 119
Ground-water pollution, 26
Grouting, 240

Infrastructure, 1, 144
Innovation, 46

Jet grouting, 240

Labor, 172
Legal factors, 6
Liability, 6

Mediation, 46
Multistory buildings, 107

Noise, 172

Quality control - management, 46

Reconstruction, 91
Regulations, 1, 6, 26
Response spectra, 181

Settlement analysis, 107
Shafts, 91
Site evaluation, 26
Soil pollution, 26
Soil stability, 144
Struts, 119
Subsurface investigations, 91

Tunnel construction, 65, 172
Tunneling, 107, 172
Tunnels, 91

Underground construction, 1, 6, 46, 65
Underpinning, 107, 240
Urban areas, 1, 6, 65, 107, 144, 172, 181, 212
Utilities, 144

Vibration, 212

AUTHOR INDEX
Page number refers to first page of paper.

Bruce, Donald A., 46

Dal Pino, John A., 107
Dowding, Charles H., 181

Finno, Richard J., 119

Gould, J. P., 144

Hahn, Daniel M., 91

Kuesel, Thomas R., 1

McCreight, Patrick, 65
Molina, Jeannette L., 6
Mooney, Joel S., 26

Nadel, Norman A., 172
New, Barry M., 212
Nicholson, Peter J., 46

Powers, J. P., 144

Rubin, Robert A., 6

Scott, David, 65
Sweeney, Bryan P., 26

Tamaro, G. J., 144
Tamaro, George, 65

Welsh, Joseph P., 240
Wyllie, Loring A., Jr., 107